eco-man

UNDER THE SIGN OF NATURE: EXPLORATIONS IN ECOCRITICISM
Editors / Michael P. Branch, SueEllen Campbell, John Tallmadge

Series Consultants / Lawrence Buell, John Elder, Scott Slovic

Series Advisory Board / Michael P. Cohen, Richard Kerridge, Gretchen Legler, Ian Marshall, Dan Peck, Jennifer Price, Kent Ryden, Rebecca Solnit, Anne Whiston Spirn, Hertha D. Sweet Wong

Eco-Man

NEW PERSPECTIVES

ON MASCULINITY

AND NATURE

EDITED BY MARK ALLISTER

UNIVERSITY OF VIRGINIA PRESS   CHARLOTTESVILLE AND LONDON

University of Virginia Press
© 2004 by the Rector and Visitors of the University of Virginia
All rights reserved
Printed in the United States of America

First published 2004

9 8 7 6 5 4 3 2

LIBRARY OF CONGRESS CATALOGING-IN-PUBLICATION DATA
   Eco-man : new perspectives on masculinity and nature / edited by Mark Allister
      p.   cm. — (Under the sign of nature)
   Includes bibliographical references.
   ISBN 0-8139-2304-2 (cloth : alk. paper) — ISBN 0-8139-2305-0 (pbk. : alk. paper)
   1. Masculinity. 2. Men—Psychology. 3. Men—Attitudes. 4. Nature—Effect of human beings on. 5. Human beings—Effect of environment on. 6. Human-animal relationships. 7. Nature and nurture. I. Allister, Mark Christopher. II. Series.
HQ1090.E29 2004
305.31—dc22   2004002631

contents

Acknowledgments   vii

Introduction   1
MARK ALLISTER

Formulating the Issues: Overviews

Deerslayer with a Degree   17
JOHN TALLMADGE

The Sky, the Earth, the Sea, the Soul   28
GRETCHEN LEGLER

"To Be a Man" in the Common Life of Nature: An Interview with Scott Russell Sanders   41
MARK ALLISTER

Chariot of the Sun: Men and the Shame of Environmental Degradation   54
THOMAS R. SMITH

Taking Care: Toward an Ecomasculinist Literary Criticism?   66
SCOTT SLOVIC

The "New Man" in Nature

Anecdote of the Car: The Diminished Thing   83
ALVIN HANDELMAN

Traversing the Timelines   98
DAVID COPLAND MORRIS

The Boys' Trip   111
RICK FAIRBANKS

"Once a Cowboy": Will James, Waddie Mitchell, and the Predicament of Riders Who Turn Writers   127
CHERYLL GLOTFELTY

Rethinking "Manly" Pursuits: Fishing, Hunting, Climbing

Fishing the Mysteries   143
  BARTON SUTTER

On the Point of a Sharp Hook   149
  JAMES BARILLA

I Love the Single Deer Path   157
  TIMOTHY YOUNG

Fathers and Sons, Trails and Mountains   168
  O. ALAN WELTZIEN

Mothers, Fathers, Sons, Wives

As Big As the World: Imagination, Kindness, and Our Little Boys   181
  JULIA MARTIN

Nature Nurturing Fathers in a World beyond Our Control   196
  PATRICK D. MURPHY

When Tillage Begins: A Family Portrait   211
  JIM HEYNEN

Husbands and Nature Lovers   219
  LILACE MELLIN GUIGNARD

Unusual Natures

Consuming Cities: Hip-Hop's Urban Wilderness and the Cult of Masculinity   235
  STEPHEN J. MEXAL

Wild Time: Prisoners and Nature   248
  KEN LAMBERTON

The Nature of My Life   257
  JAMES J. FARRELL

Notes on Contributors   275

Acknowledgments

The seed of this book was a panel on masculinity and nature for the 2001 conference of the Association for Study of Literature and Environment (ASLE). My thanks to fellow panelists Ken Lamberton, Gary Paul Nabhan, and Scott Slovic. The enthusiastic response to our panel led me to propose an anthology on the subject, and I found no shortage of willing writers. I will long remember the encouragement of ASLE members, a model academic group that is scholarly, interesting, socially committed, and fun.

Boyd Zenner, the acquisitions editor at the University of Virginia Press, has been a delight to work with. Thanks to Sue Oines at St. Olaf College and Susan Brady for their help with the preparation of the manuscript. And I am grateful to St. Olaf College for a research grant that gave me release-time from teaching.

A major project becomes bound up in a family's life, and thanks to Jan, Betsy, and Nat for not getting tired of my talk about the book. My biggest debt is to my contributors, whose ideas and prose performed that wondrous miracle of turning what I could only excitedly imagine into what I can now appreciatively see.

ECO-man

Introduction

MARK ALLISTER

Eco-Man: New Perspectives on Masculinity and Nature joins two academic fields: ecocriticism (the study of environmental literature) and men's studies. I began to see that such a joining was necessary when I taught, in the spring of 1999, an American Studies course, "Being Male in America." As I immersed myself in men's studies, I often wondered where the "nature" was, why the concerns of these academics stayed focused on city issues and human relationships—the very phrase "social constructions of masculinity" seemed to exclude nature. When I next taught environmental literature, I realized that although there was much close examination of gender, men's lives in and out of nature were discussed only from an ecofeminist position, which intertwines environmental politics with feminist principles.

This book's essays, taken as a whole, consider gender and ecology in two new ways: first, the contributors complicate the narrow view of nature as wilderness or that which is consumed by development; and second, the contributors complicate the view that sees masculinity simply as the social code of the patriarchy. When I contacted potential contributors—writers and scholars who had written about environmental literature, ecofeminism, or men's issues—I asked, "What do you have to say that's fresh about the relations between masculinity and nature?" The essays, therefore, do not present a united front, either about gender or nature. The writers in this book do not share a common theoretical position, nor do they have the same opinions about masculinity or nature. What the writers do share is a desire to make more complex the two subjects by intertwining them.

When we generalize, in our culture, about "man," we usually attribute to masculinity characteristics such as reason or intellect, elevating the mind over the body. I would argue, however, that when we are asked to name particular American men we associate with the term "masculinity," we don't usually name men who use their minds, but men who exhibit prowess in "nature," outdoors, in sports arenas, or "through" nature, by being rugged and handsome. American masculinity is notable for its emphasis on a certain conservative outward demeanor, on physical prowess, personal restraint, doing rather than thinking. And these characteristics of masculinity have come, primarily, from this country's frontier history. In his influential book *Wilderness and the American Mind*, Roderick Nash asserts the belief that "wilderness was the basic ingredient of American civilization. From the raw materials of the physical wilderness Americans built a civilization; with the idea or symbol of wilderness they sought to give that civilization identity and meaning" (xi). By the nineteenth century, wilderness in the United States was no longer seen as hideous or desolate, as the devil's lair, but as the very thing that made the country unique. Sublime scenery was everywhere, particularly as explorers, trappers, painters, and settlers pushed west. Because nineteenth-century commentators on American culture, from Alexis de Tocqueville to Frederick Jackson Turner, used wilderness as a way to claim uniqueness for the United States, they claimed that uniqueness in American character came from *men's* experience in nature. A dominant model of American manhood was created, and many famous American men, from presidents Andrew Jackson and Theodore Roosevelt to folk heroes Davy Crockett and Will Steger, exemplified that model, gaining their reputations at least in part through exploits in the wild.

This historical situation has, not surprisingly, been mirrored in literature: numerous American authors have created male characters who establish selfhood in the wilderness. In *The Machine in the Garden: Technology and the Pastoral Ideal in America*, Leo Marx notes the large number of canonical writers (Cooper, Thoreau, Melville, Faulkner, Frost, Hemingway) whose imaginations are "set in motion" by the impulse of their characters to withdraw from society into an idealized landscape. But there is no real escape, Marx suggests, looking at the numerous pivotal scenes in their novels when a machine smashes the idyll in the garden. Such scenes serve as metaphor to two dominant and competing strands of American life and Americans' consciousness: on one hand, our love of technology, new things, and efficiency; on the other hand, our love of pastoral nature. Most important to the subject of this book, however, is the connection to gender. As Marx writes, the machine is invariably associated with "crude, masculine aggressiveness in contrast with the tender, feminine, and submissive attitudes traditionally attached to the landscape" (29). The "machine in the garden" motif taps into gender paradoxes not easily resolved. For good reasons, men have traditionally been identified with the machines, from gun to plow, from bulldozer to fighter jet; and yet men have also been taught to venerate wilderness, which is usually hurt by those machines.

This paradox is particularly acute within the powerful social construction of masculinity that the way to prove one's manhood is not to test oneself in nature but to destroy it. Many boys and men fantasize about controlling big machines in the service of whacking around and altering the natural world. A cable television show consists entirely of footage of bulldozers, earth-graders, backhoes, and wrecking balls demolishing structures, moving earth, swinging steel beams. What, would we guess, is the ratio of men to women who watch such a show? What is the male fascination with these big powerful machines, which have miniature replicas in playgrounds? Perhaps the viewers long to have such power, to have such control; perhaps viewers, like children who build sand castles at the beach and then knock them down as the tide comes in, long to demolish and build. But whatever has contributed to the fascination, do we doubt that demolishing and then reconstructing the earth is not part of the social construction of American masculinity?

Social constructions of gender, as well as race or sexual orientation, are never simple or one-dimensional, and so the messages that our society conveys about masculinity and nature are paradoxical, even contradictory. Until the early 1960s, cultural critics examining literary treatments of men and nature focused typically on the "man in the wilderness" theme, and sympathetic examinations tended to emphasize the opportunities in nature for men or boys to test themselves, to experience formative activities of a kind different from those available in cities or towns. When critics considered masculinity as it was socially constructed and historically situated, the archetypal figure was usually Natty Bumppo, the hero of James Fenimore Cooper's *Leatherstocking Tales*. But from the 1960s onward, sharply critical examinations of men and nature began appearing, which argued that male self-fulfillment in the wild has come at the expense of women and children, at the expense of adult male responsibility and community. As Leslie Fiedler suggests in *Love and Death in the American Novel*, Bumppo is "the prototype of all pioneers, trappers, cowboys, and other innocently destructive children of nature" (194), and it is only a short step from Fiedler to Ursula Le Guin, who has used the term "killer stories" for those that use conflict and a linear plot to elevate the male to Hero. Jane Tompkins, in her book *West of Everything: The Inner Life of Westerns*, calls the genre a narrative of male violence whose plot hinges on men proving their courage to themselves and to the world by facing their own death (31). The Western, she argues, isn't about the encounter between civilization and the frontier: "It is about men's fear of losing their mastery, and hence their identity, both of which the Western tirelessly reinvents" (45). Tompkins's critique is easily extended to recent books not in the genre of the Western but about extreme wilderness feats, such as Joe Simpson's

Touching the Void: The Harrowing First-Person Account of One Man's Miraculous Survival or Jon Krakauer's *Into the Wild.*

Perhaps no figure of masculinity—as body and metaphor—has so completely bridged nineteenth- to twenty-first-century conceptions of masculinity as the cowboy. Beginning with the dashing heroes of Bill Cody's Wild West Shows, the cowboy has served as symbol of a "man's man." As the slogan that now appears in advertisements for Stetson Cologne suggests: What man has never been a cowboy? When Philip Morris, in 1954 the country's smallest cigarette maker, decided to change the image of its Marlboro brand cigarette, it went to a Chicago ad agency to design a campaign. The agency decided that the cowboy was the most masculine American image, and so the Marlboro Man—strong, tough, silent—"became" the cowboy and linked his version of manhood to our culture's mythic beliefs about the western United States. And Marlboro, previously a term with connotations of English aristocracy, now became the name of the country's best-selling cigarette. In her provocative book *Kill the Cowboy: A Battle of Mythology in the New West,* Sharman Apt Russell maintains that cowboys "have much to do with our cultural dreams of freedom and solitude, of riding a horse across golden fields as thunderclouds roil across the sky, of sleeping peacefully under the arc of the Milky Way, of waking alone to the bitter light of dawn. In these dreams, we test ourselves on the anvil of self-sufficiency. . . . We are intimate with animals. We are intimate with the earth" (2). In these dreams, I would add, cowboys have influenced how all Americans have thought about masculinity.

Gretel Ehrlich, in *The Solace of Open Spaces,* refines this version of the masculine code. When Ehrlich sees billboards for the Marlboro Man, she says in her chapter "About Men," the figures remind her of no one she knows in Wyoming. "In our hellbent earnestness to romanticize the cowboy," she writes, "we've ironically disesteemed his true character" (49). What Ehrlich sees and celebrates in the ranchers she knows is an "odd mixture of physical vigor and maternalism." Toughness masks tenderness: "Ranchers are midwives, hunters, nurturers, providers, and conservationists all at once" (51). She goes on to say that because "these men work with animals, not machines or numbers, because they live outside in landscapes of torrential beauty . . . because calves die in the arms that pulled others into life, because they go to the mountains as if on a pilgrimage to find out what makes a herd of elk tick, their strength is also a softness, their toughness, a rare delicacy" (52–53). Ehrlich's view of cowboys is more nuanced, in her suggesting that at least some cowboys are strong, tough, get-it-done guys, while at the same time sensitive and tender, nurturing and soft. But projecting Ehrlich's view of cowboy masculinity into the culture can be intimidating as a model of manhood.

A recent popular advertisement, which ran in numerous magazines, displays contradictions our culture teaches about masculinity and nature, because the contradictions "sell." In warm brown tones that suggest the inviting nature of the desert, not its harshness, a lone biker speeds down a newly asphalted highway through a mythic western landscape. There's no white line down the middle of the road; he has it all to himself. But this on-the-road fantasy sells a handheld palm pilot. With the palm pilot, the ad suggests, you can find your way—a box in the corner "shows" that the town of "Independence" is .20 miles up the road. You can also, as the ad's text suggests, "dream in color." Maps, photos, games, and "an ever-growing world of applications" can now be seen "in colors as vivid as your imagination."

The advertisement suggests that male success is a twining of two roles: you are a businessman who would desire or need a handheld computer; but you're not limited to that role because you are also a man-out-of-doors, a biker, tough, independent, adventurous. Magazines, newspapers, the movies, television—wherever we look in popular culture, an American man in nature, early in the twenty-first century, likely spends his weekdays working in an electrified, wired office, but on weekends or vacations he hits the road on a bike or in a boat, or competes in extreme sports, or hikes with expensive gear through stunning mountain landscapes. In any case, "nature" is a thing out there to be enjoyed, certainly not lived in. Nature is for self-congratulation: "I am still a natural man."

Just as society celebrates the masculinity on display when nature is altered or controlled, so it celebrates those men who create masculinity through and in nature. In her recent book *The Last American Man*, for example, Elizabeth Gilbert writes the life story of Eustace Conway, a North Carolina "mountain man" who lives self-sufficiently off the land—making his own clothes, hunting and gathering his food, starting fires by rubbing sticks together, bathing in icy streams, living in a tepee. Conway lives as Native Americans did centuries ago, or as a fur trapper in the American West lived in the 1820s, except that Conway doesn't wander and his goods for sale are not fur. He is not a hermit, living by himself and away from people, but a visionary, a prophet. His message is that Americans are soft and corrupt, that material goods dull the mind and spirit, that work as most people know it is empty. He believes that people can learn how to live self-sufficiently and can have ecstatic experiences in nature that are worth more than anything most Americans have now.

"The story of Eustace Conway," writes Gilbert, "*is* the story of American manhood. Shrewd, ambitious, energetic, aggressive, expansive—he stands at the end of a long and illustrious line" (127). This masculine model lives on in the enormous popularity of John Wayne, Humphrey Bogart, Clint Eastwood, or Garth Brooks. It lives on in our iconic reverence for the cowboy, and it lives on when we call successful entrepreneurs or researchers "trailblazers" or "pioneers." It lives on in our culture's put-downs of the masculinity of "sensitive New Age guys," who, in our stereotype, lack self-sufficiency and are inept at fighting or handling machines or horses. For Elizabeth Gilbert, and perhaps for many Americans, the phrase "the last American man" relates to masculinity as a hypermanhood in the wilderness. But this linking of men and nature, narrow in its conception of both nature and masculinity, speaks little to most men's lives. As I've said, it's the widening of these two—complicating nature and masculinity in relation to one another—that serves as my impulse to gather this collection of essays.

Complicating nature means that we must give up the too-easy yoking of nature and wilderness that inevitably appears when critics discuss masculinity. In his influential book *Uncommon Ground: Toward Reinventing Nature*, William Cronon notes that "nature" is not very natural, if by nature we mean that which is untouched by humans; instead, nature is "a profoundly human construction" (25). Not only have wilderness areas been significantly affected by humans, but our very notions of nature have shifted through time in response to changing psychological, sociological, and religious beliefs. There is no "nature" that is stable or foundational. And when we see that nature has numerous and shifting roles—sometimes existing as an idea, or as a commodity, or as the "pure" entity for those who want to attack what is more obviously human-made—then we are able to acknowledge how complex are the relations between this larger view of nature and masculinity. In urban settings, in backyard suburban gardens, on ball fields, as well as hiking trails and mountains—all these are arenas of nature where masculinity is described and prescribed.

The 1950s and early 1960s are often seen as a time when models of masculinity were well defined and readily accessible and needed no complicating. But bitter arguments about the Vietnam War and the questioning of that great shaper of "manhood," the military; the recognition of environmental damage; the challenges from feminists about male privilege; the increasing public awareness of gay men and gay culture—all these called into question prevailing models of masculinity. Most academic discussions in the last two decades about the socially constructed nature of masculinity have been highly theoretical, city-oriented, and focused on violence, race, and sexual orientation. Though this attention to the social constructions of masculinity has created a cultural climate that opens up the exploration of men's lives in a way that might not have been well received ten years ago, *Eco-Man* pursues a different set of issues and is far less theoretical.

If we enlarge our view of nature and examine the social construction of masculinity in relation to that view, questions that have not been asked in the past about men and nature come to the forefront. How do men involve themselves in the cycle of birth and death? How do fathers pass on environmental beliefs to children? How does gardening, farming, or ranching shape men's characters? How does experience in the natural world teach men about sorrow and loss? Can we construct a manhood around ecological principles and practices? What would such a manhood look like and how would it enrich men's studies?

In her recent book *Stiffed: The Betrayal of the American Man*, feminist writer Susan Faludi suggests that our culture ignores the "conditions" under which individual men live, reducing these men, she says, to a perennial Everyman (as women were once reduced). But how would men's problems be perceived, she writes, "if we were to consider men as the subjects of their worlds, not just its authors? What if we put aside for a time the assumption of male dominance, put away our feminist rap sheet of men's crimes and misdemeanors, or our antifeminist indictment of women's heist of male authority, and just looked at what men have experienced in the past generation?" (16). The contributors to this essay collection do such careful looking—in this case at the intertwining of men and nature.

In recent decades, sophisticated theoretical arguments about gender and ecology have come from ecofeminists, who have usefully shown the connections between environmentalism and feminism. Ecofeminists such as Carolyn Merchant, Susan Griffin, Karen J. Warren, and Ynestra King, among others, have argued convincingly that patriarchal beliefs and attitudes exploit both women and nature. Mining, logging, ranching, farming—traditional and honored ways for a man to support a family—have been attacked for the "dominance" that those acts assume is not just possible but necessary. Outdoor recreational activities of hunting and fishing are framed as testosterone-driven needs to slaughter animals. In *Ecofeminist Philosophy*, Karen Warren asserts that "there are important connections between the unjustified dominations of women, people of color, children, and the poor and the unjustified domination of nature" (1).

Though the contributors to *Eco-Man* look at gender and nature through the lens of masculinity, the book is not intended to refute the main tenets of ecofeminism. Human history suggests that humans seem nearly incapable of organizing themselves in nonhierarchical ways, and hierarchies of power tend to exploit whomever or whatever is weaker. What distinguishes the attitudes and ideas of radical ecofeminists from those of many of this book's contributors is their response to the question of whether the "patriarchy" in the United States serves as a stand-in for all or even most men and oppresses all or even most women. One example: ecofeminists are certainly justified in pointing out that it is nearly always men who bulldoze land and build houses, but the frequently made assertion that Judith Plant puts in this way in *Healing the Wounds*—"the rape of the earth, in all its forms, becomes a metaphor for the rape of woman, in all its many guises" (5)—does not take into consideration that women are implicated in buying the house in the subdivision, are implicated in our entire system of buying and selling, developing and using, and therefore implicated in the "rape of the earth."

The contributors to *Eco-Man* do not believe that gender should be placed above everything as an exclusive source of identity. Class, sexual orientation, race, religion, family upbringing, age, mental abilities, one's birth order, physical abilities, resistance to addictive behaviors—all these, and many more, shape a person's beliefs and experiences, as does gender. Men and women together are bound up in our social structures, and continuation of those structures serves some people and not others; patriarchal beliefs and attitudes exploit many women and men, as well as exploit the land and animals. Most of the essays in this book are by men, but women too are represented here. Because I wanted a collection that would represent a range of ideas and opinions about masculinity and nature, rather than a unified collection that would simply make an argument to "see it my way," I did not try to assemble a group of like-minded thinkers about the subject, and I did not ask them to produce a particular type of argumentative essay. Their voices make themselves heard in different ways, and the essays range widely in genre and approach to the subject matter.

Some readers of this book may be dismayed that there is no consistent theoretical underpinning for all the essays. Some readers may be surprised that there is no general deconstruction and denigration of the social construction of masculinity. I believe, and the book's essays taken as a whole suggest, that the social constructions of masculinity in relation to nature are a mix of good and bad, a mix that affects individual men and women, as well as our society. *Eco-Man* is intended to serve as a companion to ecofeminism, a useful addition to our understandings of nature and gender. The book, to change the metaphor, is an enterprise that trailblazes a new, but parallel, path.

If gender studies in ecocriticism have been dominated by attention to feminism, men's studies has been blind in seeing nature. The most important anthology in the field is Michael Kimmel and Michael Messner's *Men's Lives*, now into its fifth edition. This best-selling reader on men and masculinity, edited by two prominent researchers, takes the position that the "gendering process" is a central experience for men; the authors explore how working-class men, men of color, gay men, older men, and younger men construct different versions of masculinity. As one reviewer says, "This reader does a remarkable job of showing the interconnectedness of race, class, and gender."

Of the fifty-six articles in *Men's Lives*, fifteen speak directly about race, and eleven about gay men; only one article directly addresses class issues, though several others mention class or discuss the subject indirectly. Men's violence is the main subject of eight articles; violence, real or anticipated, permeates numerous other articles to the extent that violence, in addition to race and sexual orientation, permeates the anthology. This is the point: not one essay—in an anthology about men's lives—discusses men in relation to the land. There are no farmers, no ranchers, no wilderness guides. There are no men gardening or backpacking. Despite having a section on boys, the anthology has no discussion of formative experiences for boys in nature. Nothing suggests that planting vegetables or flowers, observing wildlife, wandering in woods, camping, learning to be self-sufficient outside one's house, or earning a living out-of-doors have anything to do with masculinity. For men's studies scholars, it is as if males today have spent their lives in houses, schools, and cities, exclusively, and men's "relationships" are only with humans, not the nonhuman world.

Men's Lives is not unusual in its attention to race, sexual orientation, and men's violence; I use it as example because the book is mainstream, particularly in the academic world. Most books about masculinity, no matter the politics of the author, ignore how men (and women) are shaped by, and shape, the nonhuman world. R. W. Connell's *Masculinities*, a profeminist, theoretical critique of the many masculinities in our culture, ignores the influence of nature on men. And the conservative writers advocating a sociobiologic view of gender, such as Edward O. Wilson in *Consilience*, consider nature only when they make the biological connections back to hunting and gathering communities: contemporary males' shaping by nature is largely invisible. By leaving out an important part of many men's lives—their connections to the nonhuman world—academic men's studies makes the Ivory Tower a tower from which we can see neither the forest nor the trees.

Eco-Man will look at the forests and trees, as well as prison yards, books, and malls, almost anywhere that nature is socially constructed in relation to masculinity. The book is divided into five sections. In "Formulating the Issues: Overviews," I've placed essays that address in a broad way the intertwining of masculinity and nature. John Tallmadge, in "Deerslayer with a Degree," leads off with a story of how the "man in the wilderness" model influenced his young life. But as he's become a husband and father, he writes, he's turned to a "relation and reciprocity" masculinity found not in wilderness experiences so much as in a sustainable personal ecology. Gretchen Legler, in "The Sky, the Earth, the Sea, the Soul," challenges a similar heroic model by focusing on scientist-explorers in Antarctica. "Reading" the land and the material culture left by humans, she critiques the dominant fusion of masculinity, power, and science by introducing us to the unknown women who work in that forbidding landscape. After Legler's essay I've placed my interview of Scott Russell Sanders, who has written wisely for nearly twenty years about the land and masculinity, an interest arising out of his rural boyhood and his desire to write about his life as a father. Sanders's interview, "'To Be a Man' in the Common Life of Nature," offers both an introduction to readers unfamiliar with his work and direct responses to issues of masculinity that often appear indirectly in his books. The poet Thomas R. Smith follows with an essay "Chariot of the Sun: Men and the Shame of Environmental Degradation" that is influenced by his long-standing work in the mythopoetic men's movement. Smith uses the mythic Phaethon story both to critique contemporary masculinity in its relation to the earth and to interpret our current political climate. The overview section concludes with Scott Slovic's provocative essay "Taking Care: Toward an Ecomasculinist Literary Criticism?" Using his own "ungendered" scholarship to think through the arguments of ecofeminists, Slovic offers a provisional ecomasculinism and then concludes by suggesting that all people should "concentrate on taking care, regardless of gender, to make our presence in the world as benign as possible."

In the next section of the book, "The 'New Man' in Nature," Alvin Handelman, David Copland Morris, Rick Fairbanks, and Cheryll Glotfelty all use particular geographic locales to focus on masculinity and nature. In "Anecdote of the Car: The Diminished Thing," Handelman wraps Wallace Stevens's "Anecdote of a Jar" and lines from Robert Frost poems around a wrecked car not far from his rural Vermont cabin to create a searching exploration of his evolving relations to nature. In "Traversing the Timelines," David Copland Morris, with the guiding spirits of William Wordsworth and Edward Abbey, explores the sense of loss and compensation in the mixed world of remnant wilderness and engineering colossi by using Glen Canyon Dam and Interstate 90 through Washington's Cascade Mountains as two primary sites of meditation. Rick Fairbanks, in "The Boys' Trip," uses his annual kayaking trips to the north shore of Lake Superior as a way to reflect philosophically on the relation between self and nature and to consider how different are his contemporary wilderness forays from those past masculine adventures that our culture has mythologized in classic Western films. And to conclude this section, Cheryll Glotfelty, in "'Once a Cowboy': Will James, Waddie Mitchell, and the Predicament of Riders Who Turn Writers," tells the stories of two best-selling writers who leave the ranch to make their art; she then uses their lives as models to examine how that decision changes their relationships to nature and their sense of themselves as men.

The essays in the next section, "Rethinking 'Manly Pursuits': Fishing, Hunting, Climbing," address issues that most readers might first associate with masculinity and nature. In a paired set about fishing, Barton Sutter in "Fishing the Mysteries" describes why, after a twelve-year absence, he returned to fishing, where he catches not only fish but humility, spirituality, and ecological wisdom; James Barilla explores romance and rivers, "different ways of losing the self in another," he writes in his essay "On the Point of a Sharp Hook." Just as many people wonder why anyone would want to fish—and these two essays suggest numerous reasons—many people wonder why men hunt. In "I Love the Single Deer Path," Timothy Young tells a similar story to Sutter's of quitting and then returning to the hunt, and how deer hunting has become the activity that immerses him most deeply in sacred ritual, both his own and his family's. The subject switches to climbing and wilderness hiking with O. Alan Weltzien's essay "Fathers and Sons, Trails and Mountains." Examining his own passionate desires to backpack and climb by reflecting back to his father's experiences in the mountains, Weltzien describes climbing trips he has taken with his best friends, using these trips as a way to explore male bonding and expressions of a particular masculinity.

While hunting or climbing are activities most often associated with masculinity and nature, in "Mothers, Fathers, Sons, Wives," the penultimate section, contributors reflect on masculinity from one of these positions of human relationship. The South African writer Julia Martin, in "As Big As the World: Imagination, Kindness, and Our Little Boys," writes as a mother and feminist, describing her difficulties putting her beliefs into practice as she parents Sky, her five-year-old son. Martin urges parents to teach their children to embed their lives in a particular place, an act that can enhance the child's imagination and their understanding of interdependence. Patrick D. Murphy writes about parenting his daughter in "Nature Nurturing Fathers in a World beyond Our Control." Interpreting novels, referring to environmental literature, "playing" with the differences of "to mother" and "to father," Murphy suggests that parenting a child "in nature" is a terrific way for a man to relinquish the illusion of control. In "When Tillage Begins: A Family Portrait," the fiction writer Jim Heynen explores the masculinity of his farmer father in relation to the natural world, a relation that encompasses both brute strength that attempts to conquer the world of plants and animals, and a tenderness that manifests itself in Heynen's most vivid memory, his father saving a piglet through mouth-to-mouth resuscitation. Lilace Mellin Guignard concludes this section with her essay "Husbands and Nature Lovers." She explores what are culturally considered to be alternative natures in her husband, asking questions such as: What are the forces at work in the man I've chosen to share my life with? What shaped him and how does that affect our life together as environmentalists, academics, and outdoor recreationalists?

In my final section, "Unusual Natures," I've placed essays that enlarge our sense of where masculinity and nature might intersect. In "Consuming Cities: Hip-Hop's Urban Wilderness and the Cult of Masculinity," Stephen J. Mexal discusses the inner city as metaphorical wilderness, arguing that in hip-hop's efforts to linguistically construct an urban wilderness, it invents—through performance both lyric and corporeal—a cult of masculinity contrived in response to socioeconomic disenfranchisement and performed against the metaphorical and literal backdrop of the inner city. In "Wild Time: Prisoners and Nature," Ken Lamberton looks at the nature to be found in prison yards, and at how inmates interacting with it connect to themselves and to their crimes. And in the concluding essay of the anthology, James J. Farrell's "The Nature of My Life," we return to the nature that nearly all us know best—that found in our homes, our workplaces, our malls. Farrell riffs "domesticated," "instrumental," and "ornamental" natures into a call for a new sense of ecomasculinism.

Many contributors of these essays are academics from various disciplines—literature, American studies, history, women's studies, and philosophy—but all have written here for a general reader. Seven fiction writers and poets have written personal essays for the collection. Many essays are a hybrid of scholarship and creative nonfiction. The result, I believe, is a wide-ranging, undogmatic, and eminently readable collection: a book that you might read through in order to learn more about ecocriticism and masculinity, or a collection that you could have on the nightstand, to be enjoyed one essay at a time.

Works Cited

Connell, R. W. *Masculinities*. Berkeley and Los Angeles: University of California Press, 1995.

Cooper, James Fenimore. *The Last of the Mohicans.* 1826. New York: Bantam, 1981.
Cronon, William, ed. *Uncommon Ground: Toward Reinventing Nature.* New York: Norton, 1995.
Ehrlich, Gretel. *The Solace of Open Spaces.* New York: Penguin, 1985.
Faludi, Susan. *Stiffed: The Betrayal of the American Man.* New York: Morrow, 1999.
Fiedler, Leslie. *Love and Death in the American Novel.* New York: Criterion, 1960.
Gilbert, Elizabeth. *The Last American Man.* New York: Viking, 2002.
Kimmel, Michael, and Michael Messner, eds. *Men's Lives.* 5th ed. Boston: Allyn and Bacon, 2000.
Krakauer, Jon. *Into the Wild.* New York: Doubleday, 1996.
Marx, Leo. *The Machine in the Garden: Technology and the Pastoral Ideal in America.* Oxford: Oxford University Press, 1964.
Nash, Roderick. *Wilderness and the American Mind*, 3d ed. New Haven: Yale University Press, 1982.
Plant, Judith, ed. *Healing the Wounds: The Promise of Ecofeminism.* Philadelphia: New Society, 1989.
Russell, Sharman Apt. *Kill the Cowboy: A Battle of Mythology in the New West.* Reading, Mass.: Addison-Wesley, 1993.
Simpson, Joe. *Touching the Void: The Harrowing First-Person Account of One Man's Miraculous Survival.* New York: HarperCollins, 1989.
Tompkins, Jane. *West of Everything: The Inner Life of Westerns.* Oxford: Oxford University Press, 1999.
Warren, Karen. *Ecofeminist Philosophy.* Lanham, Md.: Rowman and Littlefield, 2000.
Wilson, Edward O. *Consilience: The Unity of Knowledge.* New York: Knopf, 1998.

formulating the issues: overviews

Deerslayer with a Degree

JOHN TALLMADGE

The day we killed the rattlesnake it was beautiful in New Mexico, hot and clear with a fresh wind rippling the sage. I was hiking with a group of Boy Scouts who had come from New Jersey to experience the wild West at Philmont, a vast ranch once owned by the founder of Phillips Petroleum. We were led by a swaggering young scoutmaster who was going bald while making a good start on a paunch. He hiked at the front, followed by a gang of big boys from his own troop who lorded it over the small, skinny guys like me. I hated them with the vivid, silent hatred normally reserved for my fifteen-year-old body.

Still, it was thrilling to be two thousand miles from home, hiking along a dry, rocky trail with nothing on either side but scrub oak and yucca. No power lines, no sidewalks, no apartment buildings, no cars anywhere in sight—nothing but pure, clean air and wide open spaces. I loved the heft of my backpack stuffed with food and gear, the solid grip of my hiking boots, the sense that I could go for days and walk many miles without needing anyone's help.

Suddenly, a commotion broke out in front. The scoutmaster shouted, "Rattlesnake! Give me rocks!" The big boys were pointing at something down in the yucca, yelling and pelting away. I crowded up with the rest, clutching my own rock. But I was too late. The scoutmaster was already poking the dead snake with his staff. It was a small snake, no more than twenty inches long. He lifted it, limp and bloody as a rag, and held it aloft with a grin. The big boys cheered. The rest of us dropped our stones and fell back in line. We had killed the great serpent. The hike resumed.

I had never seen a rattler, but I had read about them in scout books. I knew they were poisonous and knew how to treat a bite. I was prepared for that. But I was not prepared for stoning a snake to death. Part of me felt a righteous glow at having made the trail safer for those who would follow; snakes were poisonous, after all. But this one had not attacked, merely challenged us with its rattle. It was a little snake. It had been overwhelmed by the mob we had suddenly become. Crushed, lynched. I was alarmed at my eagerness to take part, to grab a rock and get in my own throw. I wanted to belong, even though I really hated those guys. I wanted to feel like cheering with the rest. I wanted to feel affirmed and excited. But all I felt was disappointment and shame.

The hike went on, the weather stayed beautiful, but this trip was not turning out as I had hoped. I had been dreaming about the wilderness for years, feeding on *National Geographics* and Davy Crockett movies. Those were the days when all the young boys wore coonskin caps and made flintlocks out of broom handles. It took very little imagination to transform a cinder block wall into the ramparts of the Alamo, where Davy fought off the Mexicans, or the scrubby maples of a vacant lot into the woods where Hawkeye and Chingachgook stalked murderous Hurons to rescue the beautiful Judith Hutter.

Hawkeye, or Deerslayer, was my favorite, romantically illustrated by N. C. Wyeth in an old, frayed edition my father had owned as a boy. Deerslayer lived a wild, free, and natural life in the woods, far from the settlements with their poverty and filth. He found his own food, made his own clothing, took care of himself, and never had to answer to anyone. He was strong and capable, adept in frontier skills and woodsy lore. A great hunter, he killed with a single shot but only for food. I could not imagine him stoning a snake, or anything else. Best of all, his virtue and eloquence came not from parents, school, or church but solely from life in the woods. He was trustworthy, loyal, helpful, friendly, courteous, kind, obedient (to higher laws), cheerful, thrifty, brave, clean, and reverent (to no church but nature and nature's God). Pure and uncompromising, he always did the right thing. No wonder the dark, tempestuous Judith had fallen for him.

Of course, she was not altogether pure herself, so when she hinted at marriage, he had to decline. He had to go back to the forest where he could be free to rescue other innocents in distress. To a fifteen-year-old boy, this logic seemed impeccable. Marriage and settling down were not compatible with a heroic vocation. They interfered with your purity and siphoned off valuable energy necessary for deeds. The great thing about nature was that it provided for all your needs but never argued or lectured or went off in a huff. It never told you to pick up your underwear. You could not hurt its feelings. You could count on it in a way you could never count on people.

The best thing about Deerslayer was that he always seemed to be fully in control. He could move freely among all social classes, even crossing the color line in his friendship with Chingachgook (one of whose names, I recalled, was the Great Serpent). Deerslayer could deal with the soldiers at the fort, the settlers, the Indians, even Judith's rich and mysterious father with equal dexterity and poise. He seemed immune to seduction, despite Judith's attentions and Harry's envy. He kept his cool amid every danger or hardship. Neither his judgment nor his strength ever faltered.

In short, Deerslayer was everything that I was not, and that's what made him so appealing. He lived alone and free; I lived in a house packed with parents and siblings. He lived in the woods; I lived in the city, and not just any city, but sprawling, dingy, endless North Jersey. Deerslayer was a dead shot; I had never held a gun and hated the thought of hunting. Deerslayer was strong and graceful; I couldn't hit a baseball or sink a basket; I was more interested in science and poetry than sports. As for women, Deerslayer always stayed cool as a moose, while I was sizzling and frying with hormones. I could never, ever have turned down Judith Hutter, not that she would ever have noticed me.

Deerslayer's wilderness was attractive because it promised freedom from having to deal with women and having to compete with men. It was an arena for a kind of manhood that seemed accessible to small, self-conscious boys who lacked the killer instinct necessary for manly careers in sports, warfare, big business, or organized crime. I had hoped that the Philmont trip would put me in touch with the Deerslayer life, or at least some remnant of it, but things were not turning out that way. Instead, with the scoutmaster and his bully boys and the murder of the snake, it was beginning to look as if the home rules still applied.

In those days, Philmont was organized as a network of base camps scattered among creeks and ridges along the east side of the Sangre de Christo Range. Each camp had a resident staff and programmed activities; you hiked from one camp to another and did their activities, typical scout stuff like knots, wildlife, Indian lore, gold panning, or physical fitness. After a week of this, I was programmed out. I looked forward to free time at the end of the day when I could hunt fossils along the ridges above camp. It was quiet there, and the native plants hadn't been stomped to death.

We arrived at our last camp ahead of schedule, after running out of food when our iodine-treated water turned everything blue. That meant two days of archery and NRA hunter safety instruction, including practice with a .3006 rifle. Fed up, I stayed in the tent while the others went off to shoot. But they brought back some good news. There was an alternative to archery: we could climb a mountain just outside the ranch property. It was Baldy Peak, at 12,800 feet the third-highest peak in New Mexico. We'd have to start before dawn to make it back by dark. There were even abandoned mines and a ghost town halfway up where anyone who got tired could stop and rejoin the group on the way down.

I remember cooking breakfast by the dawn's early light, then setting off across the creek and into the trees where a fence marked the ranch boundary. On the other side, wildflowers grew knee-high, startlingly lush after the shopworn grass around camp. We hiked upstream for hours, passing beaver ponds where big trout lazed in shadows, climbing rock slides into dark spruce forests broken by aspen glades. By the time we reached the ghost town and stopped for lunch, the scoutmaster and his boys were sweating and out of breath. They decided to stay while the rest of us went on.

So it was the small but wiry guys who found the old mine tunnels and the misty meadow strewn with rusting implements, who followed the steep mining road through stunted firs to the tree line and on to the snowbank from which the creek issued, and then climbed lichened talus fields to the summit from which we could see a hundred miles in every direction. It was my first mountain and far bigger than anything in New Jersey. Our whole route was visible. We could see all the way east to the Great Plains baking in golden sun, or westward across numberless ranges mossy and green as a rumpled blanket until blue distance obscured the view. From this height every human structure shrank away into the landscape and disappeared. It looked as if the land had never been touched. It felt as if we were seeing America for the first time, as if this were the first time anyone had seen it. No wonder Deerslayer had chosen a life in the woods!

But this moment was equally precious because the little guys had prevailed over the scoutmaster and his gang, not by beating them at sports or in a fight but by overcoming our own fear and fatigue. We had made it, and our victory had deprived no one else of a chance to win. Best of all, we did not have to kill anything. Reaching that summit not only cleansed me of shame for the murder of the snake but also sealed in my mind an image of wilderness as a scene for benign yet heroic action. Here was an arena where manhood could be achieved without violence. I practically ran down that mountain. When we met the rest at the ghost town, they looked a bit sheepish even while bragging about the abandoned saloon they had found. I listened and smiled and paid it no mind. I knew I was tougher than they were.

From then on, wilderness and manhood were always linked in my mind, so naturally I looked to the woods for college. I chose Dartmouth because it was way up north and had strong outdoor traditions coupled with excellent academics. Imagine a world-class library with black bear sighted only ten miles away! Back in the 1960s, Dartmouth was an all-male school with a vigorous weekly routine: you hit the books hard for five days and then exploded on weekends. Exploding demonstrated your manhood, and there were three acceptable ways to do it: you could get laid, get drunk, or get out. There was only one way to get laid or drunk, but there were two ways to get out: you could go on a road trip (which meant getting laid or drunk), or you could go to the mountains. Needless to say, I went to the mountains a lot.

Dartmouth carefully cultivated the image of the rugged outdoorsman who was also smart and sophisticated, who smoked Gauloises and could judge a fine Beaujolais while paddling upstream and quoting Keats to a lissome damsel in the bow. Such a one would be equally at home on Wall Street or on the Allegash, a picture of total success. I loved this image! It was only years later that I realized that it was simply a greener version of what all the Ivies were selling: admission to an elite. Dartmouth's pitch was based on the model of male success that I had absorbed from the Boy Scouts and from books. The Dartmouth Man was strong, brave, smart, eloquent, self-reliant, and free of female constraints, yet unlike the citified men of Yale, Harvard, or Princeton, he was also skilled in the outdoors. He could move among all classes by virtue of a complete and well-rounded education that relied on wilderness as well as books. He was Deerslayer with a degree.

In a democracy, the true aristocrat is not the born blue-blood, but the one who can move freely among all groups or classes. Politicians like Bill Clinton have long recognized this fact; even Thoreau hints at it when he says that a man is rich in proportion to those things he can afford to let alone, or that the true walker is one who is at home everywhere. I found the Dartmouth model of manhood very affirming. It worked for me. I liked the feeling of being a cut above. If I was with intellectuals, I could talk like a woodsman, and if I was among soldiers, laborers, or woodsmen, I could think like a scholar. You could always be one thing or the other, and thus restricted to none.

When I began studying wilderness writers in graduate school, I found that most of them seemed to be following this archetype. Thoreau might have lived in the woods, but his Harvard education shielded him from what his contemporary Karl Marx called "the idiocy of rural life" as effectively as Mrs. Emerson's pies assuaged his hunger. Muir lived in Yosemite for six years as a tramp naturalist, successfully deflecting the advances of Thérèse Yelverton, countess and romance writer, while preaching his glacial gospel to all and sundry; his cabin contained books by Emerson, Gray, and Milton that he had encountered during his years at the University of Wisconsin. Abbey lived in the desert in a trailer, dreaming of mystic revelations and raw sex while commiserating with cowboys in local bars about the decline of the West; he had a master's degree in philosophy and had already published three Western novels.

These writers all lived in nature as single men with advanced degrees who found affirmation in wildness and celebrated it in books. They were moralists interested in purity of one kind or another, social critics dedicated to changing the world through writing or activism. Their words and ideas laid the foundations of the wilderness ethic that still undergirds environmental politics in America. And their vision of manhood, so deeply intertwined with these ideas about nature, was not one I seriously questioned until I reached middle age, married, and settled down to raise my kids in the city. Even today I struggle with it, for I love wilderness and remember not only my first mountain but all the others. I still love the wilderness writers and remember how their works, and the pilgrimages that they inspired, have enriched the lives of my students. But I have come to believe that we need new models of manhood if we are to achieve durable, sustainable, and honorable relations between human culture and the rest of life, especially here in North America.

What are the consequences of constructing manhood on the model of Deerslayer with a degree? Think first about the ideal of wilderness that it projects. Deerslayer's wilderness was still vast and full of game, but it was also a theater of war among native tribes and the European powers who supplied, polarized, and exploited them. It was a bloody, dangerous, and unstable place, dramatic and heroic, but also doomed by the advancing frontier. The future, even in Cooper's novel, belongs not to the virtuous and heroic woodsman but to the venal men of civilization, represented by the officers of the fort and the contemptible Hurry Harry. It is out of this world that the novel is written, nearly a century after the events it purports to describe. Cooper published *The Deerslayer* in 1841. Five years earlier, Emerson had published *Nature*, in which he had depicted the American pioneer as a type of Homeric hero. And four years later, in 1845, Thoreau set off for Walden to live a wild and literary life.

Thoreau's wilderness, like Cooper's and Emerson's, is an ideal constructed on Romantic principles, according to which virtue, strength, eloquence, and piety are absorbed through the skin from Nature itself. It is a kind of "back formation" that bears only superficial resemblance to the real, historical thing. Unlike Cooper's, however, Thoreau's wilderness has been purged of hostile natives and dangerous beasts. Even on his long trips to the Maine woods, Thoreau never encounters a bear or a wolf, and the remnant Indians have been co-opted and marginalized by Yankee culture. Violence and killing are no longer required for the woodsy life, which makes purity and virtue much easier to maintain. It is easier to have visions if you don't have to keep watching your back.

In short, the wilderness of Thoreau and his successors is a sanitized landscape. It does not menace the traveler the way the upper Missouri or the Rockies menaced Lewis and Clark. By the time Thoreau, Muir, and the rest appear on the scene, the scene itself has long been cleared of grizzlies and Indians by epidemics, hunters, and pioneers assisted from time to time by the U.S. cavalry. The wilderness celebrated by our leading environmental writers is a landscape created by the decimation of big fierce animals and recalcitrant native tribes. North America never was the place described by the Wilderness Act, where "the earth and its community of life are untrammeled by man, where man is a visitor who does not remain." Humans have lived here since the Pleistocene. The timeless, enduring wilderness with all its Edenic attributes is a dream inflicted upon the landscape by genocide. It does not correspond to reality. To deny that history is to live in a kind of bad faith.

The problem with the Deerslayer model of manhood is that the bad faith at its core compels us to repeat a pattern of repression and denial through violence. We have to keep killing the rattlesnake again and again. But whether the goal is visions or victims, wilderness always has to be someplace remote and exotic, someplace *different* from where we conduct our normal lives. Otherwise, how could woodsy lore and adventures confer distinction? You can't be a true aristocrat in a democracy, crossing class lines at will, unless the lines are there in the first place. The color line, the educational line, the line between wilderness and culture must all be maintained. This means that wilderness must be kept at a distance, shut up in national parks or forests, in order for us to idealize, worship, or benefit from it as a scene of instruction. You don't achieve manhood by camping out in the backyard.

Both these aspects of wilderness—the purged landscape, the remote Eden—have little to offer anyone who aspires to a sustainable, ongoing, ecologic relation to land. They do not invite intimacy, inhabitation, or interdependence. The landscape does not ask you to live there, do your work there, or give anything back in return for its spiritual or economic gifts. Humans are visitors who do not remain. The real wilderness man is one who passes through, an adventurer or explorer who carries off trophies in the form of knowledge, summits, stories, or even wounds, but who never assumes any burden of responsibility toward the land. Manhood becomes a matter of achievement and identity rather than of relation or reciprocity.

As for women, nature offers a poor substitute despite the eroticized rapture that runs through much male nature writing. Someone once cracked that Thoreau could get more out of two hours with a chickadee than most men could get from a night with Cleopatra. Perhaps he was right; at least the chickadee wouldn't argue, quote the classics, or embroil him in politics. Meanwhile, along comes Abbey exclaiming that he wants to possess the whole landscape of Arches National Park as one might desire and possess a beautiful woman while, in the next breath, declaring his dream of a hard and brutal mysticism in which the naked self merges with the Other and yet somehow survives intact, separate, etcetera. One can imagine the look on his lady's face.

The trouble with purity, especially male purity in this vein, is that it never changes. It doesn't lead to anything. It doesn't make room for the Other, whether animal, vegetable, or female. A manly ideal that substitutes nature for women and purity for relation offers no wisdom for a generative life, no guidance for relationships apart from a strict and formal code. Relationships tend to be messy and ambiguous, but they are also potentially fertile. Mature, adult relationships are organic and alive. They evolve: that's why they endure. They are never a sure thing, but they last as they grow. Of course they can't be "pure," but they offer the only hope for sustaining and renewing the world.

As a mature man, I remember with fondness and gratitude my own initiatory experiences in the wilderness. The trip to Philmont was only the beginning of a long sequence of journeys to places as rich, challenging, and diverse as the White Mountains, the High Sierras, the Wind Rivers, or the Boundary Waters. It is tempting, I admit, to fetishize wilderness out of nostalgia for the extravagant desires and actions of youth. But we have to remember that Deerslayer, for all his virtues, remains unfulfilled as a man. He has no home, no family, no children. He reminds us, curiously, of comic book heroes with bulging muscles and fabulous powers who keep saving the world but have no close friends and never get a date. Like Superman, Deerslayer is overdeveloped but undersexed. At some point, somehow, he will have to grow up. Or not—in which case he'll turn into a sterile, grumpy old man, Natty Bumppo of *The Prairie* who spends his time in a rocking chair dreaming bitterly about the passing of the frontier.

I believe that wilderness, even the purged wilderness of today, can still offer profound initiatory experiences to young men, especially if they are informed by a rich understanding of natural and human history. But how can we preserve such benefits while avoiding the sterility and bad faith that lie in wait for Deerslayer with a degree? Above all, we need role models for mature and generative manhood as well as for youthful heroism and purity. And we need guides for moving from one stage of life to the next. These require, I believe, a reconfiguration of attitudes toward nature, women, and the wisdom of tribal cultures. Fortunately, many of today's male nature writers have begun to explore such paths, and the results are very promising.

JOHN TALLMADGE

The Deerslayer model of manhood construes nature as a scene for heroic action. Whether the goal is meat or discovery, victory or insight, nature is set over against a protagonist who is only passing through. But today's male nature writers have begun to explore alternative ways of imagining and relating to landscape that emphasize interaction, inhabitation, and reciprocity. I call these forms of husbandry, though they embrace far more than conventional agriculture. Wendell Berry, for example, writes tellingly of farming, local economies, and the importance of community life in a vision of nature that embraces both wildness and human work. The essays of Gary Snyder and Scott Russell Sanders speak to the value of homesteading and household work carried out as part of a place-centered devotional practice. When Aldo Leopold writes of splitting and burning good oak to warm his shack, then spreading the ashes under his apple trees where, come spring, they will be transformed into fruit for both his family and the squirrels, he exemplifies how mindful interaction with the landscape can feed both body and soul. Richard Nelson, who practices subsistence hunting in Alaska, celebrates the joy of participating, consciously, in the gift economy of nature, where all life comes from other life. These forms of attentive interdependency require intimate engagement and reciprocity with the land, rather than heroic struggle or violent confrontations. They are forms of learning, the enactment (rather than the formation) of identity.

Restoration ecology is a promising trend in this new husbandry, for it seeks to cultivate not merely a handful of privileged species but an entire biota. To do this requires a great deal of knowledge, not only of individual species (many of which have no human use) but also of the ways they interact with each other in space and time. It also requires long-term devotion to a landscape. It is interesting to notice how many of today's nature writers live on recovering landscapes and engage in some form of restorational work. Replanting pine woods and prairie flora on the sand farm that he acquired for delinquent taxes brought Leopold his "meat from God." Snyder settled on a Sierra ridge near abandoned gold diggings. John Elder celebrates the return of wildness to Vermont's Green Mountains while practicing sustainable, low-impact forestry in his own hundred-acre woods. Restoration work of this kind, which requires both material and spiritual devotion, offers a powerful model for generative manhood.

Today's male nature writers also address their relations with women in ways more complex and nuanced than those of classic wilderness writers. Eros is neither projected, sublimated, nor shunned but rather is embodied and embraced. Gary Snyder's poetry is full of the earthy, organic sexuality typical of the sixties counterculture. Wendell Berry celebrates marriage as correlative to the loyal, fruitful relationships he aspires to build with land and community. Scott Russell Sanders and John Elder explore the challenges and rewards of fatherhood enacted not only in household and community but also in wilderness journeys with their children. For all these men, wives and families play a central role in their mature sense of identity and their relationships with land. Marriage and household are key metaphors in the vision of a sustainable, personal ecology.

Finally, contemporary nature writers have recuperated indigenous wisdom and values to a degree not seen since Thoreau's writings on Maine. In the century after Thoreau, American nature writing tended to ignore or caricature Native American cultures. Gary Snyder was one of the first environmental writers to recognize, learn from, and celebrate Native American myths and customs; he envisions a "future primitivism" that blends the "Old Ways" of tribal peoples with the wisdom of modern science and scholarship. Barry Lopez's encounters with Arctic people reveal the resourcefulness, dignity, strength, and virtue that indigenous life ways can foster. Some of his best stories depict how respectful attention to the land and its creatures can produce mental health by bringing the individual's "inner landscape" into accord with the order and ecological coherence of the "outer landscape."

Other writers have explored ways to participate more directly in native culture. Richard Nelson began his career as an anthropologist conducting ethnographic studies of Inupiat, Kutchin, and Koyukon hunters. He spent many years living and studying with tribal elders, learning the techniques and rituals of hunting and eventually coming to appreciate viscerally their spiritual sense of the landscape as living and aware. Nelson's own practice of subsistence hunting draws heavily on the wisdom of his native teachers but also on the ecological perspectives of Western science. It is an ingenious syncretism that, while developed and practiced in Alaska, could provide a model for personal ecology elsewhere.

Another writer involved with indigenous cultures is ethnobotanist Gary Paul Nabhan, who has spent years working with dryland farmers in the Sonoran desert. Nabhan has devoted himself to understanding native foods and traditional farming practices, disseminating his research results to all resident cultures. He leads seed prospecting expeditions to native gardens, both current and abandoned, and he takes people on "food pilgrimages" where they rely on a local, traditional diet and then, at the end, ritually stomp on boxes of modern foods like white bread and potato chips that have been implicated in high rates of diabetes and heart disease. This work not only provides a rich harvest of stories but also empowers native communities while identifying valuable new crops (the jojoba bean is perhaps the best-known example). One might call such a practice "restoration anthropology."

The version of manhood celebrated by these writers in their poems, essays, and stories has much in common with the old archetype of Deerslayer with a degree, yet it also has key differences that point toward generativity. Like Cooper's hero, these men are all physically strong, adept in the outdoors, and full of woodsy lore. They aspire to a principled life, and they are gifted with words and stories. On the other hand, they have all made a commitment to inhabit the landscape, not merely to pass through it on the way to adventures; they live in place as householders and citizens. They relate meaningfully and honorably to women. They maintain marriages and families, and they are involved with children. Many of them engage in teaching and activism, as if determined to give something back. Many live on recovering land and are working at restoration. Many are involved in working with and learning from native people. They are in touch with wildness, not only through heroic journeys but also through attentive engagement with the modest, local landscapes in which most of them actually live. They offer a model of mature manhood as a practice of reconciliation, where traditional antagonisms (human or wild, native or alien, male or female) are finally overcome.

Faced with a rattler, it's hard to imagine any of these men casting the first stone.

THE SKY, THE EARTH, THE SEA, THE SOUL

GRETCHEN LEGLER

It was blowing so hard outside that the outhouse was rocking. Inside, there was brief relief from the wind and some relief, although not much, from the bitter cold. I had driven out on snowmobile with Kendrick Taylor to this place, the Siple Dome drillsite on the West Antarctic Ice Sheet. Taylor, a scientist from the Desert Research Institute in Reno, Nevada, and the principal investigator for this project, was inspecting preparations for the season's work, a pulling up of miles of thin, round cores of ice from this ancient frozen slab, in hope of finding out something about the history of the Antarctic continent and current trends in global warming.

Aside from the outhouse, the only two other places to get away from the howling wind and freezing cold were the drillsite bar—named Little Alaska and housed in a musty, dark, Korean War–era, wood and canvas jamesway—and the eerie, blue, silent ice trench, a long underground tunnel under the frozen snow where the salvaged ice cores would be stored to "settle" for a year before they were shipped to laboratories around the world for study.

It was 1997, and I was in Antarctica as a guest of the National Science Foundation's Artists and Writers Program, exercising the great privilege to visit McMurdo Station, the main U.S. base in Antarctica, as well as the South Pole, numerous scientific field camps, weather facilities, and fuel and supply stations scattered across the continent. My writerly mission was simple—to go and see—to see with my own eyes the solid earth of this once mythical land, *Terra Australis Incognita*, and to craft what stories I could of my travels in it: stories of the land's complexities, its people, its history and politics, and its beauty, but mostly stories of how the land came to be what it was, shaped by wind, by snow, by time and by men.

I was in the outhouse not to get out of the wind and cold but to take a pee. Stripping off layer after layer of cold weather gear took me some time, and as I was engaged in disrobing I started to read the walls around me, remembering with a smile that I had been told to look for this outhouse, as it was rather famous in Antarctica, traveling from field camp to field camp, gathering more signatures and messages at every stop. This outhouse was the incongruous place where—like the stone walls along canoe routes in the American Northwest that hold the ochre handprints and signs of moose and bear left by aboriginal peoples—year after year, scientists, drillers, other support workers, and visitors could leave their mark upon Antarctica, engraving their passage this way in the world. Irreverently, I mused as I sent a stream of dehydrated, dark yellow pee down into the depths of the pristine ice below me, that this outhouse gave us humans a chance to mark our trail, not unlike the way a dog marks its passage down a neighborhood street, pissing out the borders of its territory. The outhouse, and all the other buildings at Siple Dome, including the galley jamesway with its wooden Café Bubba marquee, the jamesways used for sleeping, for bathing, for the camp medic, and the one for the scientists' office space, would be dismantled at the conclusion of this particular science project and, if needed, moved somewhere else, say, from the West Antarctic Ice Sheet to the Shackleton Glacier, where the same huts would house a different team of scientists and support personnel. This moveable feast of an outhouse was famous for its graffiti, which liberally covered its walls, in both neat and scrawling script, letters large and small, conveying messages profound and absurd. As I sat, I read what was in front of me on the wall:

Come, my friends, 'Tis not too late to seek a newer world.
Push off, and sitting well in order smite the sounding furrows; for
   My purpose holds to sail beyond the sunset, and the paths of all
Western stars, until I die. It may be that the gulfs will wash us
Down: It may be we shall touch the Happy Isles, and see the great
Achilles, whom we knew. Tho' much is taken, much abides; and
   Tho' we are not now what strength which in old days moved
   earth
And heaven; that which we are, we are; One equal temper of heroic
Hearts, made weak by time and fate, but strong in will, to strive, to
Seek, to find and not to yield.

These are the unmistakable words of the wandering hero Ulysses, as written by the poet Tennyson. They are the words inscribed at the base of the statue of Robert Falcon Scott, whose bronze figure stands in a pretty square by the river Avon in Christchurch, New Zealand. Scott was the British naval officer who just missed being the first man to stand at the South Pole—losing the honor to Norwegian Roald Amundsen in 1911—in an epic journey that epitomized the manly British quest for fame and honor for one's country, made all the more glorious as Scott and his men died in the effort. These are the words invoked by countless explorers and adventurers in Antarctica, as in any new land that humans have tried to subdue: to strive, to seek, to find, and not to yield. No matter that this quintessentially masculine call to arms most times melts under close scrutiny; what one is supposed to strive for, seek, and find is at best vague and at worst indefinable: Honor, Truth, Virtue, Glory. What are those? And what is it to which one is not supposed to yield? Snow, Wind, Ice, Thirst, Fatigue, Frostbite—one's own Body, Nature?

Antarctica is a land made by men, a tabula rasa like no other, except perhaps the dark blankness of outer space. It is a clean, white, icy slate upon which has been written a code of masculinity so romantic, so bold, so nostalgic, that it made me, sitting there in the outhouse, want to cry and laugh at the same time. The masculinist tradition established in Antarctica by its early male denizens, first those historic whalers, and then the explorers of the Age of Heroes, certainly is evident today. There are still men who work in Antarctica who wish woman never had set foot there. There are still male scientists who resent the attention being paid to their female colleagues, and who hang tenaciously to the antiquated ideology of the first modern scientists, who see Antarctica as a woman, a virgin, her body lying in ready to be penetrated, probed, and dominated, her veiled secrets revealed. While in Antarctica, I overheard an eminent volcanologist one day refer to Ross Island's 12,000-foot-high volcano, Mount Erebus, as "Antarctica's last untouched virgin."

This masculinist ideology still holds mostly unexamined sway, and no wonder; it comes from the very heart of the enterprises that have characterized human activity in Antarctica since the continent was first sighted in the early 1800s. Those activities were: the whaling industry with all its inherent savage capitalism; early exclusively male expeditions fueled by Victorian-era nationalism and imperialism; the establishment of an all-male military presence in Antarctica during the Cold War years; and finally, in 1997, the transferring of control of the continent's U.S. interests from the U.S. Navy to the National Science Foundation. One can clearly see, in this pedigree, the makings of a masculine landscape.

From where I sat in the Siple Dome drillsite outhouse, my eyes continued to roam the weathered plywood walls, encountering more neatly written excerpts from heroic literature. "Polar exploration is at once the cleanest and most isolated way of having a bad time which has been devised." These words come from the introduction to *The Worst Journey in the World*, an account of the Scott expedition by Apsley Cherry-Garrard. "Cherry" also covered in his account the journey from Scott's Cape Evans hut to Cape Crozier, on the other side of Ross Island, where he and two other expedition members, Edward Wilson and "Birdie" Bowers, wanted to collect newly laid emperor penguin eggs. The emperors lay their eggs in the dead of the Antarctic winter. Wilson, the expedition scientist, had had this in mind when he first joined Scott's party—to work out the embryology of the emperor penguin, what he thought was probably the most primitive bird in existence. The expedition nearly killed them. When they returned, finally, after bitter nights in unbelievable storms, having had to sleep in reindeer hide sleeping bags frozen stiff with ice, their teeth split into pieces by the cold, having had to defecate in their trousers and shake the frozen shit loose, they were hardly recognizable as the men who had left five weeks earlier. Scott wrote in his diary of their return: "To me, and to everyone who has remained here, the result of this effort is the appeal it makes to our imaginations, as one of the most gallant stories in Polar History. That men should wander forth in the depth of a Polar night to face the most dismal cold and the fiercest gales in darkness is something new, that they should have persisted in this effort in spite of every adversity for five full weeks is heroic. It makes a tale for our generation which I hope may not be lost in the telling."

Scholars of Antarctica have remarked that for many of the early explorers, and no doubt countless more contemporary ones, Antarctica was and is a state of mind, a state of imagination. Scott put it so plainly: "The result of this effort is the appeal it makes to our imaginations." Antarctica, then, was and is still a stage for our human fantasies of nobility, honor, and masculinity. That Wilson, Cherry, and Birdie had performed their journey for science made it all the nobler. "The British," wrote Antarctic historian Roland Huntford, "had an exalted view of science as moral uplift." *Dulce et decorum est pro scientia mori* (It is sweet and proper to die in the name of science), Antarctic historian and marine scientist Donal Manahan put it one evening during a presentation in McMurdo Station's galley, part of the Sunday Science Lecture series that had been part of the routine for Scott and his men and that is still a part of McMurdo operations almost a hundred years later. The line between heroism and foolishness sometimes seems rather thin. Had they any business going out into the teeth of the Antarctic winter night for penguin eggs? But what if they had never tried, what then?

Perhaps the answer to that question comes from another message on the outhouse wall, from Arctic explorer Fritjof Nansen:

> People perhaps still exist who believe it is of no importance to explore the unknown polar regions. This, of course, shows ignorance. It is hardly necessary to mention here of what scientific import it is that these regions should be thoroughly explored. The history of the human race is a continual struggle from darkness toward light. It is therefore, to no purpose to discuss the uses of

knowledge. Man wants to know, and when he ceases to do so, he is no longer a man.

I had a feeling that this contribution must have come from a scientist, in response to such manly kidding as was revealed in this quip scribbled nearby: "Keep Antarctica Beautiful. Nuke a Beaker," "beaker" being the word that was used by nonscientists to refer, derogatorily, to scientists in Antarctica. Or perhaps "Nuke a Beaker" was not the chicken but the egg, being written in response to the assertion that intellectual acuity and imagination were the manliest of traits and that not to have them, not to pursue them, meant one was, well, a wimp. I felt, reading the outhouse walls, that I was witnessing the story-writing mechanism of Western culture at its most primitive.

What impressed me equally as much as the way these early explorers' narratives had taken hold of the contemporary imagination and sunk themselves deep into the core of the myth of Antarctica was the similarly powerful narrative of science in this land. Scientists had become the new heroes of Antarctica and its new story writers, whose tales drew from the tales of the explorers before them, creating a complex, layered narrative.

One story about science in Antarctica will never leave me. It was told to me by Rice University professor Rob Dunbar, principal investigator of the ROAVERRS project, which stands for Research on Ocean Atmosphere Variability and Ecosystem Response in the Ross Sea. He told me the story one day as the icebreaker we were on, the *Nathaniel B. Palmer*, made its way through the waters of the Ross Sea on its research mission. At the time the story took place, Dunbar said, Japanese and Russian fishermen wanted to harvest krill from the Antarctic Ocean for human and animal consumption. Krill are the small shrimplike creatures that live in massive clouds in the southern oceans. The would-be harvesters reasoned that since baleen whales were now mostly dead due to overfishing, the krill, which was once their chief food, would go to waste if not harvested for humans. But wait, others cautioned, there was evidence to suggest that the crabeater seal had now become entirely dependent on krill for food. But no one knew for sure, so a study was launched.

The sampling tools for the study included automatic weapons and knives. The scientists doing the fieldwork would get set down on an ice floe with automatic weapons, kill a group of seals, slash them open, take their ovaries and their teeth (markers of sexual maturity and age), and then return to the ship covered in blood. What they found out was that female seals were becoming sexually mature at earlier ages, at from between two and three years of age, as opposed to an earlier five or six. It was a rapid and surprising response to a surplus of krill; an evolutionary change that took only a few generations to leave its mark. The seals had indeed filled the whales' niche and had become the primary eaters of krill.

"I don't know of what value the story is," Dunbar concluded, "perhaps it's only of value for the tale that it tells." Echoes from Scott's journal came floating back—"That men should wander forth in the depth of a Polar night. . . . It makes a tale for our generation which I hope may not be lost in the telling." What tale did Dunbar's story tell? Was Dunbar one of the scientists who slaughtered the seals? Was the tale his way of confessing that he, personally, felt uncomfortable about science's legacy of brutality against the natural world? Was there a way to do science without resorting to this kind of muscular violence? Was Dunbar's tale, like Scott's story of gallantry, a tale to appeal to our imaginations, and if so, in what ways?

On the outhouse wall, near the Amundsen quote about masculinity, power, and science, I read this: "And now there came both mist and snow, And it grew wondrous cold: And ice, mast-high, came floating by, As green as emerald." It was part of Samuel Taylor Coleridge's "The Rime of the Ancient Mariner," in which the mariner's ship gets blown off track, ending up in the horrible ice-choked Antarctic seas, and the albatross comes to guide the ship to safer waters; out of boredom the mariner shoots the bird, and the ship once again is beset in devilishly calm seas, the dead bird being hung around the mariner's neck as punishment. Next to Coleridge in black magic marker was an excerpt from Percy Bysshe Shelley's "The Cold Earth": "The cold earth slept below / Above the cold sky shone, / And all around / With a chilling sound / From caves of ice and fields of snow / The breath of night like death did flow / Beneath the sinking moon."

Besides the poetry, the wall was also festooned with adolescent sex jokes, present in outhouses and bathroom stalls all over the world. "Would you, could you, on a tower? I would, I could, I did." Near that was drawn a cutaway picture of a huge ice drill, like the one towering outside, and the hole it was making in the ice sheet. The depths were numbered and labeled: 150 meters, coreplay; 250 meters, intercore; 430, coretus interuptus; 507, drill stuck, need giant tube of core jelly; 554.19, coregasm.

In Little Alaska, the drillsite tavern next door to the outhouse, a chest-high plywood bar served as the place where the drillers rested their grimy, Carhart-clad elbows at the end of the day to sip their Danish bitters or beer. It was no proper tavern, but a dirty, cold, green canvas hut, smelling of diesel fuel and cigarette smoke, the wind buffeting its thick plastic windows. The wood on the bar was carved and painted with jokes and quotes, just as in the outhouse, some nonsensical, some bizarre.

"Maybe I'd sell you a chicken with poison interlaced with the meat." Or, "Find'em, Fuck'em, and Forget'em." Also carved deeply into the wood, with pen tips, screwdrivers, and pocket knives, was a long list of answers, made over time by many hands, to the question, "How do we survive in this world?" *Sarcasm. Racism. Feminism. Patriotism. Environmentalism. Industrialism. Territorialism. Nepotism. Ethnocentrism. Anthropocentrism. Cynicism. Alcoholism. Isolationism. Orgasm. Paganism. Hedonism. Nudism. Anarchism. Sexism. Rugged Individualism.*

The metaphorical possibilities of the outhouse and the deeply engraved drillsite bar are irresistible—different expressions of a code of beliefs and values, written on the continent, carved deeply into the wood of it, into the ice and rock of it, a set of values that at the dawn of the twenty-first century seem as if they should be moderated, perhaps even outdated, both for their heroic absurdity and for their macho barbarism. But here, in the Great White South, the Last Great Continent, the Unknown Southern Land, they still play themselves out with a nineteenth-century urgency. The potent brew of masculine values, imperialism, and military ideology that were a part of Antarctica's history effectively created an indelible ink with which the story of Antarctica has been written—a history full of men's tales of bravery and heroism, folly, foresight and imagination, but a history, all the same, that has another side—the story of women in Antarctica.

It wasn't until 1974 that a woman was allowed to stay the winter at a U.S. station in Antarctica. Up until that time the U.S. Navy would not allow women, even scientists, to set forth upon the continent because of such insurmountable difficulties as not having separate bathroom facilities for men and women and the problem, considered grave, of the danger of a woman being raped by lonely, sex-starved men.

"There are some things women don't do. They don't become Pope or President or go down to the Antarctic," seemed to be the overwhelming sentiment, voiced in 1947 by Harry Darlington, as he tried to dissuade his new bride, Jennie Darlington, from accompanying him on the Ronne Antarctic expedition, a journey Jennie writes about in her book *My Antarctic Honeymoon*. Darlington is also famous for his response to the question, "What do you miss most in Antarctica? Fresh food?" No, he said, scandalously, not even women, but "temptation." Admiral Richard E. Byrd imagined Antarctica as a sacred, femaleless place, free of temptation, a monastery, a Christian holy land: "The whole of Antarctica might be referred to as a mighty cathedral of glittering ice and painted sky erected by the Lord's own hand. Far from the turmoil and temptations of the world, it is the ideal retreat for those who find a more intimate touch with the infinite greatness and goodness."

Not only were women forbidden to set foot in Antarctica, preserving the continent as all-male space, but the language and images used to represent Antarctica have until recently been thoroughly masculine, colored not only by the exclusively male colonization of the continent but also by the early association of the military with Antarctic exploration. Early book titles included *Assault on Antarctica; Assault on the Unknown; Strong Men South; The Siege of the South Pole*—works that spur the imagination to conjure the landscape as a kind of hell, the scientists as heroes, the soldiers as lone knights.

While men had been busy attacking, laying siege to, assaulting, pitting their wills against, being strong in and strengthened by, dominating, proving their patriotism in, and getting themselves frostbitten and killed in Antarctica, as well as writing all about it, what were women doing? From what you might be able to read about women in Antarctica, you would be led to only one answer: nothing. The popular *Reader's Digest* coffee table book *Antarctica: The Extraordinary History of Man's Conquest of the Frozen Continent* claims to be the complete story of this unique place. For all but two of its 320 pages, however, it leaves discussion of women out, except in the chapter titled "The Women They Left Behind," a short essay about the unsung wives of the Antarctic explorers.

As Virginia Woolf asked the women students at Newnham and Girton Colleges to whom she spoke in her book *A Room of One's Own:* Women, why have you been so lazy? Why are the works of Shakespeare not by you? Today one might ask: Women, while all the men have been so busy leaving their marks in the frozen, lonely South, where have you been? Have you no heroism in your blood? Why is the South Pole station not named after you? Have you not wanted to die for science or for your country? Have you not wanted to go and see what you could see?

The answer is, Yes; women *have* wanted to go to Antarctica, they have had adventurism in their blood, just as they have had it in their blood to travel the globe as men do, to even, in our time, enter space. But they found it difficult to get to Antarctica. As Barbara Land, author of *The New Explorers*, put it: "Women had no part in these early expeditions. They were left at home for more than a hundred years." There is a female history of Antarctica, but it has been lying low, as women's history often does, unnoticed, unwritten, unread, unremarked upon in the bigger book of human history that often highlights men's supposedly more colorful and dramatic achievements.

The first woman ever to set foot in Antarctica is said to have been Caroline Mikkelsen, the wife of a Norwegian whaling captain. On 20 February 1935, she is said to have stepped ashore (no doubt in a voluminous dress) for a short time near the present location of Australia's Davis Station on the Antarctic Peninsula. After she looked around, she was taken back to the ship. The next women to get anywhere near the continent were the wife, daughter, and two female friends of shipping magnate Lars Christiansen. The women never got off the ship, but nevertheless the landmark Four Ladies Bank, just off the Ingrid Christiansen Coast, was named after them.

Edith Ronne and Jennie Darlington were next in the late 1940s. They accompanied their husbands to Stonington Island near the Antarctic Peninsula. They spent a year in Antarctica as part of an ill-fated expedition, becoming the first women ever to live on the continent.

The first woman ever to do scientific research on the continent was a Russian marine geologist named Marie V. Kenova, who had worked in the polar regions for nearly thirty years before she got her chance to actually go ashore in 1956.

While the U.S. Navy continued to stall with lame excuses about bathroom facilities, other countries were letting women work and live on Antarctic bases. Four Argentinean professors, for instance, did hydraulic research on the continent in 1968 and 1969. Their first seemed to open a door for others, as 1969 saw, finally, the first U.S. National Science Foundation–funded all-female research team, headed by geochemist Lois Jones from Ohio State University. Jones had originally resigned herself to doing all her research about Dry Valleys lakes from her lab in Ohio, as, because of her gender, she was officially banned from the continent. When she and her research team did arrive, they were paraded around by the navy as part of a publicity stunt—even sent to the South Pole to meet navy brass and have their photographs taken: the first women *ever* to set foot on the ice at the South Pole. The headlines that accompanied the stories declared, "Powderpuff explorers invade South Pole." Reporters asked questions about lipstick, hair, and the difficulty of personal hygiene in such a harsh environment.

Dr. Mary Alice McWhinnie, an American expert on krill, like Jones, did research in Antarctica for ten years, cruising the oceans just off the continent but never setting foot on land, on forbidden territory. In 1974, she became the first woman ever to head an Antarctic research station, having been appointed the chief science officer at McMurdo, where, in another first, she spent the Antarctic winter, along with her required "assistant," Sister Mary Odile Cahoon, a biologist and nun. The two women have become mythic—the middle-aged scientist and the nun—two maiden aunts, asexual, motherly, and nurturing. Perhaps those in charge reasoned they were the perfect women to set the stage for others to follow—no men would want to rape them (too old, too nunnish), they wouldn't disturb the sexual dynamics of the base, and they could provide motherly and sisterly companionship.

Other women followed, breaking down barriers: in 1979, Michele Rainey became the first woman selected to be the year-round doctor at the South Pole; in 1993, the first all-women's expedition reached the Pole on foot from the Weddell Sea, and in 2001, a team made up of Minnesota schoolteacher and polar explorer Ann Bancroft (also the leader of the 1993 all-women's expedition) and Norwegian teacher Liv Arneson crossed the entire continent on foot.

In time, the navy's insistence on keeping women out of Antarctica became something of a curiosity. Why all that trouble to keep women out? The answer is both simple and complex: to preserve a space, a womanless space, a clean, pure, celibate, priestly space, a space like an icebound Eden before the Fall, for the uninterrupted production and reproduction of all that we know of as masculine, so that heroic expeditions like Scott's, like Amundsen's, and others that came after, could be reenacted, endlessly, thereby protecting something elemental from change, from time, from history—what it means to be a man, and by elaboration, what it means to be a human being. At stake was nothing less than the definition, albeit an extremely limited one, of our very being—*Who* are we?

In the austral summer of 1997, unofficial reports suggested that of the one thousand and some-odd carpenters, plumbers, electricians, scientists, laboratory technicians, administrators, secretaries, janitors, cooks, pilots, firefighters, doctors, dentists, nurses, mountaineers, heavy equipment operators, recreation specialists, and others working at McMurdo Station, about 40 percent were female, up from near zero percent three decades earlier. Informal interviews revealed that most men at McMurdo, even those who had worked in Antarctica twenty seasons, who had been present at the beginning of the revolution, thought that having men and women at McMurdo normalized the town and made it a better place to work and live. "Without women around," said one clean-cut male engineer, "men are pigs." It all brings to mind the Stephen Crane short story "The Bride Comes to Yellow Sky," about the marshal's bride who, amid much controversy, comes to live with her new husband in a turn-of-the-century frontier town and, in the end, exerts a civilizing influence on the Wild West outpost.

For me, to be a woman and a lesbian in Antarctica was not so unlike being a woman and a lesbian at home in America. As always, I felt I was somehow disturbing a delicate equation, except that this feeling was lessened greatly by the comfort of what seemed a larger than normal population of sisters. There was a running joke at McMurdo: "How do you get a date with a woman in Antarctica?" Answer: "Be one." Antarctica was a place, because of its remoteness and its requirements for self-sufficiency, strength, and self-possession, that seemed to attract a large number of independent-minded, strong-bodied women, both heterosexual and lesbian. Wouldn't it be wonderful, I often mused, not to think of Antarctica as a masculine space but to rewrite its myths to tell a tale about this exciting outlaw population: Antarctica as home to a female Amazon race of bulldozer drivers, glacier pilots, women scientists, and intrepid female explorers? What would it be like to take the whole of the story that had been written so far and wipe the slate clean and start anew? Perhaps instead of becoming an icy version of Herland, however, one could go a step further and Antarctica could become a land where gender didn't matter, where the rules of the Antarctic Treaty that prohibit national claims on any Antarctic lands and insist on use of the land for the purposes of peaceful scientific endeavor in perpetuity could also insist that this great magical land also remain gender-neutral. It could be the place where a new race of androgynous beings dwelled, those beings Virginia Woolf so admired who were possessed of what she called man-womanly or woman-manly minds—incandescent minds. Antarctica would be the perfect place to realize this incandescence Woolf wrote of—a place where the actual, concrete difference of gender could be revealed as irrelevant in the face of a land where the human is so small, and the earth is so large.

The outhouse was still rocking with wind, still bitterly chilling, the seat frosty, the walls glistening with frozen crystals. I was in the process of redressing. But, wait, there was an empty space on the wall. I too could leave my mark! What would I say, what would I pass down for all to read? Shelley? Tennyson? "The Ancient Mariner"? Scott? What came to mind instead were the words of the narrator in Ursula K. Le Guin's short story "Sur," the tale of a fictional group of South American women who ventured to the South Pole in 1909. The women reach the Pole (two years before Amundsen and Scott) but decide to leave no sign, no mark of their being there—no rock cairn, no flag, no pictures, no tent, no message to their king. They know this will mean that their trek will go unnoted, unrecorded, ignored. They know this and do not care. They did not go for science. They did not go for their country. They did not go to leave a mark. They went "simply to go and see. A simple ambition, I think," the narrator says, "and essentially a modest one." At the Pole the narrator writes that there seemed no reason to leave a mark: "Anything we could do, anything we were, was insignificant in that awful place." The story ends ironically with the birth of a child at the expedition base camp at the Bay of Whales, where the expedition members are safely picked up by their ship, the *Yelcho*. The child is named Rosa del Sur, the Compass Rose, and delivered without incident in the snug icy cave the women had carved for themselves. In the end, the narrator says of their decision to leave no mark at the Pole, "I was glad then that we left no sign there, for some man longing to be first might come some day and find it, and know then what a fool he had been, and break his heart." The "backside of heroism," says Le Guin's narrator, is "often rather sad; women and servants know that. They also know that the heroism may be no less real for that. But achievement is smaller than men think. What is large is the sky, the earth, the sea, the soul."

I left no mark on the outhouse wall. In fact, Antarctica may bear no imprint of me at all (except maybe my frozen pee—will it come up five hundred years from now in some scientist's ice core?), but the place itself will have marked me, seared itself like an icy hot brand on my soul. I zipped up my parka and stepped out into the cold, sweeping, flat white of the ice. I, like Le Guin's heroines, had discovered that in a place like Antarctica, stripped so bare of the accoutrements of human cultural life, a place so close to the "chaos and old night" that Thoreau spoke of from his ecstatic perch on the top of Mount Katahdin, the rules of masculinity and femininity reveal themselves as the flimsy constructs I have always known them to be. The only serious difference between men and women in an environment this inhospitable is how they urinate—men can reveal only the tiniest piece of flesh to the raw winds, while women must practically disrobe and put a square foot of bare skin at risk of frostbite. But even this difference has been challenged: pink plastic penislike funnels are issued to women who want them, so they too can pee standing up.

With the usual distractions of modern Western life pared to a minimum in Antarctica, the workings of the machine of culture become more clearly visible. I am reminded of the gift a friend of mine in Antarctica received from a pal of hers in Idaho to keep my friend happily occupied over the long Antarctica winter. It was a see-through engine kit, the engine casing clear plastic and the guts of the machine in multiple colors, so that you could, when it was all glued together, see the mysterious intricacies of the machine in motion. In Antarctica, one could see how this piece of neutral geography—hunks of rock, ice, and snow—could become not only a gendered landscape but a hypermasculine one. I could see clearly how certain values, ideologies, myths, and stories could be applied, like so much paint, to the rocks, ice, and ocean, so that the place took on a mantel of human value. And how the mythos is created is not mysterious in the least. Men "made" Antarctica, land of ice and snow, home of the blizzard, hell on earth, the great white South. "Constructed" is a better-understood word for this kind of myth making these days. Because, really, evolution made Antarctica; time made Antarctica. Ice and snow and volcanoes and wind made Antarctica. No human made Antarctica. Man is just as unnatural in that environment as woman. And despite the stories we create on the slate of its flatness, Antarctica will always remain simply what it is.

"TO BE A MAN" IN THE COMMON LIFE OF NATURE
An Interview with Scott Russell Sanders

MARK ALLISTER

Scott Russell Sanders began his prolific career—he's published over twenty books—writing environmental fiction and children's books. With the appearance of *The Paradise of Bombs* in 1987, he established himself as one of the best writers of creative nonfiction in the country. That book has been followed by five more, including *Staying Put; Hunting for Hope;* and *The Force of Spirit.* When considering his childhood in rural Ohio or his roles as father and son, Sanders has written often about masculinity and nature.

As to the interview's title, Sanders writes: "In our common life we may find the strength not merely to carry on in the face of the world's bad news, but to resist cruelty and waste. I speak of it as common because it is ordinary, because we make it together, because it binds us through time to the rest of humanity and through our bodies to the rest of nature" ("The Common Life," in *Writing from the Center*).

MA (Mark Allister): In your introduction to *The Paradise of Bombs*, in 1987, you write that you need to puzzle over "what it means to be a 'man' because the spectacle of women waking to their own full powers has pushed—or should have pushed—all men to such puzzling." Do you still, in 2002, feel the need to puzzle over what it means to be a man?

SRS (Scott Russell Sanders): When I wrote those lines, my son was eight or nine years old, just beginning to notice how grown-up men behaved. Jesse was paying attention to male athletes, rock stars, actors, teachers, soldiers, even politicians. I thought a lot about the confusion of models, good and bad, that he would encounter as he tried to imagine what sort of man he wished to become. At the same time, my daughter, Eva, was entering adolescence, and I wondered how she would find her way in a world still dominated by men. I didn't worry as much about the kinds of female models she would follow, because she had grown up in an era and a household that celebrated women. Jesse, on the other hand, was bound to hear countless challenges to maleness, as I had in my own boyhood. He'd hear that men were too violent, selfish, insensitive, fickle, boastful, and competitive, that men were obsessed with power, that men exploited women and animals and the earth, that men were the source of most evils in the world. How could a boy sort through all those charges to discover what is worthy in manhood? I wrote *The Paradise of Bombs* in part to answer that question for myself, to ponder the masculine models that had shaped my own identity. I also wrote the book to figure out how to speak of masculinity with my son. I'm still puzzling over what manhood means or ought to mean. I don't see how any man who is alert to our times can escape such questioning.

MA: Could you say more about the masculine models that shaped your identity—what you learned about these models by writing the essays?

SRS: When I was growing up, the men I looked up to were mainly neighbors in rural Ohio and heroes in books. I didn't go to movies, didn't listen to pop music, rarely watched television, so my horizons were local and literary. I admired farmers, carpenters, masons, mechanics, electricians, horse-trainers—men who worked with their hands, who told stories with a rough sort of eloquence, who understood animals and machines. I saw them mainly outdoors, or in workshops and barns. Most of them hunted and fished. They had shotguns leaning behind their kitchen doors. Many of them built or fixed up their own houses, repaired their own cars. Some of them were in and out of work. I was also fascinated by characters like Huckleberry Finn and Tom Sawyer—not yet men, but knockabout boys who'd been forced to grow up fast. I read about space travelers, inventors, explorers—men who took risks to make discoveries. None of these men, real or literary, was much given to introspection. None of them was what you'd call intellectual. The only exceptions to this were teachers and preachers, a few of whom made deep impressions on me—as I've written in *The Country of Language*. After my father died, grief pushed me into writing a series of essays about him, and only then did I realize how profoundly he'd shaped my sense of manhood. Those same essays helped me see what masculine models I still believed in, and which ones—such as soldiers and swaggering athletes—I should discard. *The Paradise of Bombs* was the most important book for me in working through that mixed inheritance, especially the title essay and "The Inheritance of Tools" and "The Men We Carry in Our Minds."

MA: I know you've written at length about your father, but could you suggest some of the ways in which he shaped your sense of manhood?

SRS: He never talked about his feelings, and he rarely let them show. He'd show anger, especially when he'd been drinking, but most of the time he radiated a musing confidence and contentment, a sense of enjoying life and being in control. He never showed grief or fear or longing, never admitted being vulnerable. He never cared much about art or books or music, but he loved everything in nature, the crops in the fields as much as the trees in the woods and the fish in the creeks. He came from the country, never had much money, and wasn't much impressed by the things money can buy. He didn't care about the appearances of places or people—about the outsides, the costumes, the facades—but he probed the interiors with amazing insight. He looked after people—us children, my mother, neighbors, folks in the church, friends at work. His relatives lived hundreds of miles away, in the Deep South, but whenever we visited them, he looked after them, too. He used to carry tools in the car, never knowing when he'd need to fix something. He was tremendously skillful with his hands, could repair anything we owned. He liked using his body. He was proud of his strength. That's a sampling.

MA: In a recent essay, "The Uses of Muscle," you write about how physical strength mattered to your father and the other men you knew in childhood. What are some of the connections for you between working men, rural life, and the muscles that you as a boy lacked and wanted?

SRS: The men I knew on the back roads of Ohio worked with their hands and backs. They dug postholes and ran fence, they shoed horses, they pitched silage, they overhauled engines, they laid concrete blocks, they cut down trees and split logs for wood stoves. I wanted the strength and the skill to do that kind of work. So I was always pushing at the edge of what I could really handle. I loaded hay bales onto wagons, and then from wagons into barns, coming home at night with my forearms scratched raw and my back aflame. I carried lumber and mixed cement and climbed onto roofs with bundles of shingles over my shoulder, until I was so tired I couldn't lift an arm to comb my hair. Although I was a scrawny kid, I would never admit I was too small for a job, and I wouldn't quit. Gradually I learned how to do many of the things my father and the neighbor men knew how to do, and I was proud of that. Then I went off to an Ivy League college and read modernist literature and learned that all those men, their strength, and their skills were held in pretty much universal contempt by sophisticated people. At first I was bewildered, and then I grew angry. I'm still troubled by the way manual labor tends to be scorned in our high-tech, brainy society. I've never lost my admiration for those rough, handy, rural men, including my father—which is why they keep showing up in my writing.

MA: "The Men We Carry in Our Minds" is an eloquent testament to the lives of these "rough, handy men." From their examples, you say in the essay, you grew up believing that men's destinies were to be either "warriors or toilers." When did you begin to think that your life might be neither of these, might veer in a far different direction?

SRS: What set me thinking about a life of the mind was science. My father and his buddies knew how machines worked, how electricity or water ran through a house, how to deliver a colt or a calf. I admired their knowledge. Maybe I got a little drunk on it, because I wanted to understand the workings not just of machines and houses and animals but the whole universe. So I read everything I could get my hands on about science. In grade school I collected fossils, did chemistry experiments in the basement, looked at stars through a cheap telescope. By high school I was building model rockets and lying awake at night working math problems in my head. The path to a career in science lay through college, so I went off to Brown intent on studying physics. By the end of my sophomore year, however, I had been seduced away by literature. My divorce from physics was complicated and painful, and it had a lot to do with disgust over the harnessing of science by the war machine in Vietnam. I've told a version of that story in *Secrets of the Universe*. Although I never became a scientist, the dream of understanding the whole shebang—life, consciousness, the cosmos—set me on a path to become a writer, an outcome that thoroughly puzzled my father.

MA: Did you and your father argue about your change of plans?

SRS: We argued about the Vietnam War, until I persuaded him that I had thought through my objections carefully—that I wasn't just protesting because lots of other young people were. And then he respected my decision to become a conscientious objector. After all, he'd taught me to think for myself. We never argued about my switch from physics to literature. It just puzzled him, that's all, and I suppose it made him wonder how I'd ever earn a living.

MA: Would your father have been puzzled if you had become a physicist, someone who used almost exclusively the brain rather than the body? Why, in particular, was he puzzled by your desire to make sense of "the whole shebang"?

SRS: My father had a great native intelligence, and he was quick with numbers—he'd won a statewide prize in mathematics as a schoolboy—but he didn't see how a man could support a family by putting marks on paper. He grew up on a hardscrabble farm in Mississippi, came of age during the Great Depression, spent a year in college on a boxing scholarship, then gave up books and caught a bus to the bright lights of Chicago. From that point on, all his learning took place on the job, on the farm, or in the streets. The notion of his son becoming a physicist probably struck him as more sensible than the notion of his son becoming a novelist or poet. At least scientists made things happen in the world. The desire to understand the universe was all right for a physicist, he figured, because at least their theories could be tested by experiment. Writers, on the other hand, invented things out of whole cloth. My father was a practical man. I never heard him speculate about the big questions—the origins of the universe, the meaning of life, the foundations of morality, the sources of evil, the reasons for beauty, and so on. He couldn't look at a machine without trying to figure out what it was for, how it was made, how it worked, but, so far as I could tell, he didn't have the same curiosity about the cosmos.

MA: In your essay "Buckeye," you write about a remarkable moment after your father's death when you saw him as a red-tailed hawk. Could you tell us more about this experience, and what in your view of things encourages and doesn't encourage you to acknowledge such intuitions?

SRS: As I say in the essay, I found myself returning to the little farm we'd owned in Ohio, years after most of the land had been flooded by a reservoir, and years after my father had died. It was a November day, with the first iron cold of winter, and so I bundled up before climbing out of the car. The land was grown up in brush and trees where we'd had pastures and gardens and barns. I couldn't bear the thought that all our work on that place had been erased. Eventually I noticed two huge weeping willows, their crowns rising higher than the power lines. And I remembered planting those willows with my father, when they were little slips no bigger around than my finger. Now the trunks were as thick as my waist. I took off my gloves and laid my hands on one of the trees, and suddenly I began sobbing. My whole body shook. I hadn't cried that hard since the day I heard the news of his death, and I haven't cried that hard since. I wandered off into the woods, bawling. When I came to where the reservoir had drowned the trees, a red-tailed hawk gave a sharp cry and launched out from a branch over my head, and began circling. I looked up and saw absolutely that the bird was my father. It was a red-tailed hawk, and it was my father at the same time. I didn't just sense this vaguely; I *knew* it as sure as I was standing there. Immediately I quit sobbing, I felt calm, I felt at peace. It was like a switch thrown in my heart. I didn't hear any words, didn't receive any mystical insight. I just understood something new about death and life.

MA: Had you glimpsed reincarnation? The transmigration of souls?

SRS: I wouldn't use those terms. I'm also wary of talking about God. But I sensed a *continuity* between my father when he was alive and this soaring bird. There was an unbreakable thread binding these two creatures—maybe there was a web binding all creatures. I realized that everything my father did, every action and thought, keeps rippling out through my life and through the lives of everybody he touched and every place where he visited or worked.

MA: Did this new understanding come all at once?

SRS: The *seeing* was instantaneous. The *understanding* came slowly.

The drive home from our old farm in Ohio took me six hours, and all the way I thought about what had happened. When I got home I made notes about it in my journal. But I didn't tell anybody, and didn't think of writing about it for publication, because I knew how easy it would be to explain away. Psychologists would call it projection, a sign that I hadn't dealt with grief over my father's death. In light of the scientific worldview, which I had studied and embraced, it could only have been a hallucination, a mirage, a delusion. But as I lived with the experience, let it settle in, I couldn't shake the *reality* of it. It was as convincing as anything I'd ever experienced. And I realized that much of what really matters to me—loving my wife and children, loving the land and its creatures, caring for my students, feeling loyalty to place and friends and art, feeling joy from everyday blessings—could be explained away by the same logic as the work of selfish genes or hormones or evolutionary conditioning. Unless I was willing to discount all of my central experiences, I had to accept that I really had seen my father as a red-tailed hawk. So finally I wrote about it in "Buckeye," which appears in *Secrets of the Universe*, and I returned to it in "Hawk Rising," which appears in *The Force of Spirit*. Since publication of the earlier essay, I've heard from fifty or sixty people who've had similar experiences. Either we're all crazy in the same way—which is probably what the psychologists would say—or the world is subtler than a materialist philosophy allows for. I've come to suspect, in fact, that what we call matter is a manifestation of consciousness or spirit, rather than the other way around. After I read "Buckeye" in public for the first time, a Shawnee man from the audience told me that the experience I'd found so mystifying made perfect sense to him in light of the vision handed down among his people.

MA: American culture, particularly academic culture, has typically separated science and spirituality, but you've written that "the geography of land and the geography of spirit . . . are one terrain." Do you think that your intertwining of all modes of perception, an opening up to the world that might be similar to that of the Shawnee man in your audience, makes possible such experiences as your seeing your father as a red-tailed hawk?

SRS: Because you use the word "perception," I can't help but think of that riveting line from Blake: "If the doors of perception were cleansed, the world would appear to man as it is—infinite." Not only infinite, I would add, but *one*, a single unfolding reality. We parse that unity into bits with language and numbers and concepts. We talk about "spirit" and "matter," "culture" and "nature," "I" and "you," quarks and kangaroos and Oedipal complexes. And of course that dividing-up of reality is handy; it enables us to talk and think about the universe instead of merely dwelling here; it gives rise to art and religion as well as science.

The trouble comes when we take the imaginary lines we draw on the universe for real divisions. Academics and religious zealots commonly mistake their theories for the reality they're trying to understand. In our society, men may be especially prone to this confusion, obsessed as we are with carving up the planet into scraps of property, thinking we can reshape the earth to our own designs. One aim of spiritual practice, in many different traditions, is to free us from this delusion and to restore us to a sense of the undivided whole. My own glimpses of this truth have been fleeting, but they have also been utterly convincing. Many of them have come in the presence of the world that humans have not made—red-tailed hawks, sycamore trees, humpbacked whales, amoebas, maidenhair ferns, gnats, swallowtail butterflies, thunderstorms, pebbles, creeks. When I think back through my books, I recall struggling again and again to re-create those moments when the unity of things shines through. Of all my experiences, those seem to me the ones most worth recording. But of course it's impossible to do them justice. Language balks and breaks down.

MA: I'd like to follow this line a bit longer but turn the subject more directly to gender. One of my favorite Blakean aphorisms is this: "Man has closed himself up, till he sees all things through the narrow chinks of his cavern." Blake was most likely referring to humankind, but I'm always struck when I read Blake how prescient he is about many men's lives right now. Do you see contemporary cultural constructions of masculinity opening things up for men—cleansing the doors of perception—or closing them down?

SRS: I think the masculine models available in American culture are more varied now than ever before. Clearly, there are still the muscular, aggressive, brash men who star in violent films and football games. They appeal especially to adolescent boys, and to some women who like the whiff of testosterone. Soldiers are back in vogue, especially since the terrorist attacks of September 11, 2001. You can see ads for them on television, scaling walls and slaying dragons and stalking faceless enemies. Male politicians still get mileage out of posing with chain saws and quarter horses, and some of the older ones still prove their virility by marrying women a generation younger. At the other extreme, at least on the testosterone scale, there are the "sensitive" men, including some celebrities, who enjoy cooking, gardening, design, meditation. They may study Buddhism or some other spiritual tradition, including the mystical strain in Christianity. They may stay home to care for children while the mothers go off to work. They may practice some form of art. And of course, in virtually every profession, there are now gay men who make no secret of their homosexuality, and who challenge many of our inherited notions about masculinity. In between these extremes—between the poles of force and feeling, if you will—there are countless gradations, offering men far more models. While plenty of men still get stuck in the male-as-muscle, male-as-suit, male-as-killer roles, I'd say that on the whole men are moving toward a liberation from the old emotional and perceptual constraints suggested by the quotation from Blake. Some of that easing of constraints is thanks to women, who for a generation now have been pushing men to rethink the meaning of masculinity. Some of it is thanks to the heady mixing of cultures—which gives us the Dalai Lama, for example, as an antidote to John Wayne, or Salman Rushdie as an antidote to the Ayatollah Khomeini, or Nelson Mandela as an antidote to Saddam Hussein. Men are still willing to retreat into caves and stare out through narrow chinks, but we don't *have* to; we can stay outside, gaze up at the stars.

MA: Your use of the Dalai Lama, Salman Rushdie, and Nelson Mandela as counterexamples to traditional masculinity are all, I note, men from non-American or non-European cultures. They also are intellectuals and writers. But when I consider education and American boys—particularly how a boy learns about masculinity from sources other than the family—I usually think that popular culture has the greatest influence. Would you agree with that?

SRS: Alas, I think you're right. I've read that by the time the average child graduates from high school, he or she has spent more hours watching television than attending school. And hour for hour, television no doubt leaves a deeper imprint on the young imagination than do all those lessons in school. This means, in effect, that for most boys, TV *is* school. If you add in the hours spent listening to pop music, going to movies, looking at comic books or magazines, and playing video games, then it's clear that many boys surround themselves with popular culture nearly every waking hour. Of course, "popular culture" is not a monolithic phenomenon. It doesn't peddle one and only one vision of masculinity. But as a first approximation of the truth, I would say that *much* of the television, film, and music popular with young people tells boys a single story: The purpose of life is to grab for yourself as many toys and sensations as possible, to acquire power over others, to get sexual pleasure from girls, to be violent and swaggering, to spend money, to be cool—which means buying the latest products—and to sneer at any serious use of the mind. That may sound harsh, but I think it fits an awful lot of what young people see and hear and read these days. In an essay called "Death Games," from *The Paradise of Bombs*, I've speculated on how boys in particular might be affected by this message. In a traditional society, a boy coming of age would be taught how to do the necessary work—hunting, fishing, trapping, farming, building shelters, caring for animals. He'd be told, "Because you are no longer a child, you can no longer merely seek your own pleasure. Now you are responsible for the welfare of other people." In our society, a boy coming of age hears from the media, "Have fun!" In other words, "Stay a child." Only now the boy has a strong body, an alert penis, a car, a charge card, maybe a gun, and it's dangerous to keep encouraging him to act with a child's selfishness and hedonism.

MA: In your last two responses you've gone from sunny to gloomy—sunny outlook, in the models of manhood available to men; gloomy outlook, in what preteen and teenage boys are "encouraged" by our culture to be. If we follow this line, two questions emerge for me: One, how do we change the dominant images that constrict boys? And two, how do we help boys, as they move toward adulthood, choose something other than the male roles typically shown in television and the movies?

SRS: In the long run and on the largest scale, we need to transform our culture. In place of pop media that sell stuff by exploiting narcissism, violence, cheap sex, and hedonism, we need to foster a culture that addresses our capacity for imagination, creativity, affection, curiosity, and compassion. Such a change will only come about as part of a larger transformation of our society from one based on money and consumption to one based on conservation, simplicity, and a regard for the common good. That's an enormous task, of course, and to many people it might seem an impossible one. However long the odds, though, I'd rather work for a more humane culture than simply drift along with the one we have. One way or another, it's going to disappear—either because it will self-destruct or because we'll create something better. Every one of us can help in that effort—through the books we write, through our teaching, through the way we go about our jobs, through what we say over the supper table or at the high school gym, through our work in the community. In our own households and neighborhoods, we need to offer our boys alternatives to the electronic media. We can start by turning off the TV. We can take our boys outdoors—to play sports, take hikes, go canoeing, walk around the block. We can give them real work to do, work that matters, such as cultivating vegetables in the backyard, painting the shutters, or hauling trash out of vacant lots. We can get them involved in volunteer work in the community. We can share our skills with them—sewing clothes, building furniture, drawing pictures, playing the guitar, fixing cars or bicycles, carving wood, doing plumbing and electrical wiring. And if we don't have any skills to pass on, we can learn some. Above all, we need to spend time with our boys—talking, listening, working, sharing food and books, playing games, fixing things. The responsibility for ushering boys into the adult world is one that all men share, but especially those of us who are fathers. No amount of toys, clothes, cars, or private lessons will make up for an absent father.

MA: The impetus for writing your recent book, *Hunting for Hope*, was an argument you had with your son, Jesse, about the push and pull between what our culture promotes as the good life and what that good life, collectively, does to the natural world. You explore in great detail what in nature humans might turn to that will get us beyond the despair that comes when we look around and see what we're doing to the environment, and each other. Could you explain what you discovered as sources of hope?

SRS: During Jesse's teenage years, he and I quarreled in the way that sons and fathers often do. We argued over the usual things—cars, clothes, money, music, girls. He kept demanding freedom, and I kept demanding responsibility. All the while, I sensed there was a deeper grievance between us. Finally, on a Rocky Mountain backpacking trip that we took the summer before his eighteenth birthday, in the aftermath of a furious argument, he told me what pained him most deeply was that I had denied him hope. He said he'd been taught to see problems on all sides—war, racism, shabbiness in the media, corruption in business and politics, devastation of nature—but he hadn't been given an equally powerful vision of solutions. He'd learned to see so much darkness in the world, he'd lost track of the light. Although I took his accusation personally, he wasn't blaming me alone; he was blaming my whole generation. I couldn't erase the problems or pretend not to see them. Nor could I offer neat solutions. I couldn't show the path to nuclear disarmament or peace in the Middle East. I couldn't say how to reverse global warming or cure AIDS. But I could say what I put up against all those enormous challenges. I could tell my son, my daughter, my students, and my readers where I find hope. So I spent the next three years writing *Hunting for Hope*, which is an inventory of sources of healing and renewal. I wrote about the creative power of wildness in nature, including our own bodies. I wrote about the benign side of human skill and knowledge, from welding and surgery to bread making and astrophysics. I wrote about our capacity for kindness, compassion, and fidelity. I looked at family and community as training grounds that can nurture our best qualities. I celebrated beauty as a revelation of harmony between ourselves and the universe. I set forth my convictions about the way of things—the grain of the world, the ground of being—and about how we might align our lives more fully with that transcendent pattern or source. All of this was spread over a couple of hundred pages, mind you, so I can't re-create the whole inventory of hope in these few words.

MA: How did Jesse feel about *Hunting for Hope*?

SRS: He was my first reader, because in a way he was the hero of the book, and I wanted to make sure, before I published anything, that he felt all right about how I had presented him. His first response was to be amazed by how seriously I had treated his concerns. I hadn't just let them blow past, but had brooded on them, had tried to understand and then answer them. The book gave him courage, he told me. It helped orient him, helped him see a way forward.

MA: Do the sources of hope that you found and articulated in *Hunting for Hope* connect particularly to masculinity, or do you think that the same sources of hope can also comfort and motivate women?

SRS: I certainly hope—there's that indispensable word again—I hope the book speaks to women as well as men. It was prompted as much by questions from my daughter and my female students as by questions from my male students or by the challenge from my son. But perhaps some of what I say in *Hunting for Hope* will strike male and female readers differently. Soon after the book was published, for example, I received a letter from a woman who said she'd never read anything by a man that dwelt so fully on relations between parents and children, or that spoke so affectionately about the body as a source of knowledge and delight. I believe it's true that men—at least in America right now—tend to think less about the fate of children than women do, and that men are more inclined to think of their bodies as weapons or tools. And these differences in attitudes about our bodies are linked to divergent views of nature. Women are more likely to recognize their bonds with the rest of life and with the earth, if only because they experience the cycles of wildness in their own bodies every month, and because many of them bear and care for children. Men are just as fully natural, of course, but they're more likely to imagine themselves apart, triumphing over nature with money or machines. Maybe the clearest challenge to the reigning images of masculinity offered by *Hunting for Hope* is the fact that a man wrote it, and that so much of what I celebrate in the book is either familial, ecological, or spiritual. I don't talk about money or machines or politics as sources of hope. I don't talk about rolling up our sleeves and getting under the hood and fixing whatever's broken in our civilization. I don't list bullet points for action. I don't propose declaring war on despair—the way male politicians have declared war on poverty, drugs, and terrorism. I don't blame women, and I don't idolize them. True, the chapter on "Beauty" opens with an account of my daughter's wedding, as seen through the doting eyes of the father of the bride, and that may be an example of the notorious male gaze. But "Beauty" goes on to speak about the photographs beamed down from the Hubble Space Telescope, about music, books, mathematics, birds, and other gorgeous designs, all suggesting that our ability to see and relish these patterns tells us something profound about the universe and ourselves.

MA: You see "gorgeous designs," beauty, in innumerable places that our materialist culture never suggests that we look, perhaps in part because the looking is "free"—it doesn't cost money. Men are bombarded in American popular culture with messages about sex, particularly in relation to buying goods or projecting an image that depends on the man spending much money. Can experiences in the natural world be an antidote to cultural sex messages? You've written that nature is sensuous if we learn to look at it carefully—what do you mean by that?

SRS: Our bodies are made to delight in the planet! Turn healthy children loose outdoors and watch them taste and listen, watch them stare, watch them sniff the air. They exult in the natural world. Our kids would be better off if we let them spend more time playing in vacant lots or turning over rocks in creeks, instead of sitting in front of screens or rushing from homework to lessons. We grown-ups would be better off if we took walks after supper and simply opened ourselves to the world. Our senses, our muscles, our emotions, our language—all evolved over thousands of generations in intimate contact with the earth. The moon and stars wheel around inside us, men as well as women. The wind blows through us, the sound of rain pours through us. Our technological society cuts us off from that contact, shuts us up in buildings, in cars, in cyberspace. It starves our senses and our souls, which hunger for the earth. Then the hucksters exploit this craving to sell us junk food, sleazy television, electronic toys, sport utility vehicles, trendy clothes, Caribbean cruises, you name it. Of course advertisers use sex to sell practically everything—especially to men, because we respond to sexual stimuli more quickly and crudely than women do. Sex is a great and mysterious force. Turning it into a commodity is not only offensive, it's dangerous. Traditional societies know that. They surround sex with rituals, ceremonies, myths, and taboos because they realize that this generative force can tear apart households and communities. Our commercial culture has belittled sex, trotting it out like a pet dog to do tricks. The epidemic of rape, unwanted pregnancies, broken marriages, sexual abuse, jealous rages, and predation in our society reminds us that sex is no pet dog. This is an untamable power. Only a people who had lost a sense of the holiness and magnificence of nature would try turning sex into a toy for peddling stuff. Only men who've lost touch with the earth would fall for shabby sales pitches.

MA: The nature writer Paul Gruchow has said that his family and hometown friends thought that he should have been a pastor; those who are passionate about environmental issues often write forcefully and eloquently about cultural and social problems, speaking, metaphorically, as if they were in the pulpit. How do you strike a balance in your writing between the urge to testify or prophesy and the urge to show, to move people by story?

SRS: I try to strike a balance, but sometimes I fall off the fence. On one side of the fence is pure storytelling; on the other side is pure sermon. The best writing, whether about nature or masculinity or anything else, gives us memorable stories and lively images, but it also tells us something important about the world and about ourselves. I try not to preach, because I'm well aware of my ignorance. At the same time, I want to be honest about what I see and think. So I often wind up speaking against the dominant culture. I wind up saying that much of what's called "progress" in America impoverishes our way of life and damages the earth. I say that money is not the most important measure of wealth, and that the consumerist nirvana projected by television is shallow and empty. I say we're foolish to worship technology or to regard science as the only way of knowing. I say other species have inherent value, apart from any use we might make of them. If you merely echo what most everybody else—from the White House to the nearest billboard—is saying, then nobody raises an eyebrow. But if you speak against that consensus, you're liable to be called preachy. Your voice stands out. I've never set out to be contrary. I've never imagined myself a prophet. I've tried to be honest. I've tried to be a good witness. I think that's crucial work for art—to bear true witness, to see and show the world in a new light.

MA: You've written in various genres—fiction, the essay, children's literature—but what unites this work is your tilt toward the natural world and toward environmental issues. Given what you say, above, about speaking against the dominant culture, in what ways does the label "nature writer" fit you, and in what ways does it not?

SRS: If being a "nature writer" means acknowledging that humans are animals, that we live in a web of relationships with other species, that our welfare depends on the health of the planet—then I'm a nature writer. But the label implies that I've chosen some special field, like sports writing or travel writing, whereas I'm merely paying attention to the most elementary facts about our existence. It's like being called an "air writer" because you notice that people breathe. We are *made* of nature. Our bodies, our senses, our minds, our language are all shaped by nature. They draw their vitality from their wildness. So the fact is I don't think of myself as a nature writer. I think of myself as a writer who remembers what our ancestors have known for thousands of years, which is that the earth is our precious and irreplaceable home. I don't just write about ferns and waterfalls and red-tailed hawks. I write about people and the things people do. I write about weddings and funerals, about doing science and baking bread, about the stuff of daily life. But I never forget that our lives flow out of the wild world and back into it again.

Chariot of the Sun
Men and the Shame of Environmental Degradation

THOMAS R. SMITH

The Fall of 2000

At the time of the 2000 presidential election, I found myself becoming preoccupied with the threat of environmental disasters likely to follow a George W. Bush victory. Surely my awareness of the very real possibility of an eventual ecological catastrophe served to salt the political wounds of that election. Almost every man is, to borrow psychologist and men's movement leader Robert Moore's expression, "hard-wired" to protect his family, of whomever or whatever that may consist. In our time, instantaneous global media have, for better or worse, knit us all—human and nonhuman alike—into a single diverse (and, one is tempted to say, unhappy) family. And yet men have in large part failed to protect the earth, our vast extended family's only home. Since 11 September 2001, the Bush administration has sought to further divert us with an apparently endless series of wars from our most crucial task of survival. For the majority of men, I believe this deflection can mean only increasing shame, whether experienced consciously or sublimated as apathy, depression, rage.

In the fall of 2000, I personally experienced shame for my part in the degradation of Mother Earth. That shame hardly lessened in the interval between the ill-fated presidential election and the Republican assumption of legislative control in the 2002 midterm elections. For those of us who love and wish to preserve the planet, watching Bush's unilateral withdrawal from the Kyoto Accord, aggressive efforts to exploit wilderness areas for oil drilling, and dismantling of environmental protection standards has entailed severe emotional and spiritual suffering.

During that dismaying winter in which the Supreme Court enthroned a president who had received fewer popular votes than his opponent, I happened to read the myth of Phaethon in Ovid's *Metamorphoses*, a story uncannily descriptive of the dangers we face in our current crisis. For weeks I became obsessed with the story, which, as all authentic myth does, speaks to us fully in the present despite its ancient provenance. It is the genius of such stories that they refuse a single fixed interpretation and instead offer themselves to the reader by way of any number of possible approaches.

This passionate immersion in the Phaethon story brought sharply to mind a period in my life, not far back in years but light years away in shifting cultural and political moods, when I, along with tens of thousands of other American men, made an earnest, if by mainstream standards eccentric, study of mythology and poetry under the aegis of what was known in the late eighties and early nineties as the "mythopoetic" men's movement.

Traditional mythology always contains at its core the question of human beings' right relationship with natural and cosmic forces, often personified as gods, demons, and spirits of various kinds. There are probably sound correlations to be drawn between our modern loss of mythology and the utter contempt for nature displayed by our cowboy-minded national leadership. I like to believe that in investigating mythology a dozen or fifteen years ago, the men involved in the "mythopoetic" movement were also strengthening bonds of loyalty to the earth and cosmos, on which the mythological imagination has suckled from time immemorial.

Initially the story of Phaethon fascinated me for its immediate relevance to our twenty-first-century environmental emergency. To briefly summarize, young Phaethon asks his father Phoebus Apollo to prove his paternal love by allowing Phaethon to drive the chariot of the sun, a task the father performs daily as light-bringer to the world. The father has promised the son anything he asks, so he can only try to dissuade Phaethon from his dangerous wish. If Phaethon backs down from his request, there will be no story; we know that this escapade will not end happily. Predictably, the boy loses control of the solar chariot, which swerves too near the earth, boils the oceans to steam, and sets mountains on fire. In answer to Earth's pleading, Jove halts Phaethon's inadvertently destructive spree by striking him dead; thus Phaethon's innocently conceived excursion concludes, but only after extreme damage to the living environment.

These images powerfully evoke the projections of a dry, overheated planet that climatologists have been making for decades. As I studied the story, I became aware of more subtle layers of meaning embedded in its catastrophic sequence. Specifically, I found myself thinking more and more about the odd relationship between the naïve son and the father who permits the son's self-destructive behavior. I began to speculate about the inner, or—as the archetypal psychologists call it—*intrapsychic* meaning of the story, especially its implications for men in relating respectfully to the natural world. The ancients saw the general male possibility encapsulated in the pairing of the Puer, the flighty, untempered boy, and the Senex, the wise, earth-knowing elder. American men, at least by evidence of recent governmental policies toward the environment, are a Puer-dominated group. Could it be, I asked myself, that how a man respects the earth depends at least in part on the extent to which he has balanced the influence of the Puer and the Senex within himself?

In the spirit of the mythopoetic men's work circa 1990, I decided to examine this Greek myth, bringing to it the sort of associative play typical in the numerous mythopoetic men's groups of that time. In doing so, I hope to demonstrate this myth's continuing value to men (and women as well) struggling against the political currents of the day to protect our threatened environment.

The Mythopoetic Moment

Before turning to take a closer look at the Phaethon story, let's dwell for a moment on the philosophical and political background of the mythological men's movement.

The psychologist Robert Sardello writes in *Freeing the Soul from Fear* that the archetypal images the soul produces "are not what we 'see'; they are what we 'see through.' . . . We experience a sense of soul when we feel the sense of these deeper patterns working through our sensing, thinking, feeling, and acting" (xx). Much of the actual substance of the men's mythopoetic work in its heyday lay in identifying the enduring mythological patterns manifested in the lives of contemporary men. This process of identification was not so much a matter of inventing "new myths" as of locating one's personal experience within a greater archetypal or mythological landscape. The basic act of naming the mythological dimension of our lives allows us to more effectively assert individual destiny against the large, impersonal fates that we usually unknowingly serve. If the mythic theme strengthens life, our consciousness of it will allow us more freedom in choosing its fulfillment; if the mythic content is destructive, and we can see that clearly, we may be better empowered to alter our course.

Robert Bly's best-selling discussion of a Grimm Brothers tale, *Iron John: A Book about Men* (1990), set the tone for much of the mythopoetic men's work, justly winning widespread fame for its author. Bly's colleague, the Seattle storyteller and drummer Michael Meade, in similar style illuminated stories from various world traditions in his unjustly neglected *Men and the Water of Life* (1993). In fact, these two books constitute most of the limited literary flowering of what has remained chiefly an oral tradition, still carried on to this day in a few tenaciously rooted men's mythopoetic enclaves around the country, long after the mainstream dismissed them as self-indulgent frivolity.

I joined the Minnesota Men's Council in the mid-1980s, following the first Bly-led men's conference in our region. For the better part of a decade, forty or fifty men met monthly to explore with persistence and intensity of purpose the whole subject of maleness, which many of us had given scant consideration in our eagerness to embrace feminism and feminine values during the 1970s. Some participants in our group had admitted feeling uncomfortable confiding in men; by cultivating a capacity for intimacy with each other *as men*, therefore, we broke new emotional, psychological, and spiritual ground.

The typical method in such groups involved creative group analysis of ancient stories from varied world traditions, an approach modeled by Marie-Louise von Franz, the Jungian thinker whose method Bly adopted in *Iron John*. Unfortunately, small associations of men scattered around the country applying Jungian concepts such as "shadow" and "anima" to their own lives weren't very newsworthy; distorting Bly's metaphorically subtle "Wild Man" figure in *Iron John*, the mass media resorted to portraying caricatural "wild men" running naked in the woods.

In fact, the American right wing employed its full resources of media and advertising persuasion to ridicule and diminish the serious attempts some men were making, however unconventionally, to live more conscious and soulful lives. (Bly likes to cite as an example of this belittlement a sleazy Dewar's ad reading, "You don't need to beat a drum or hug a tree to be a man.") The reactionary men in power in the United States had, and still have, a vested interest in keeping men, especially younger men, unconscious and brutal in the interests of maintaining American military and economic dominance.

In addition, some feminists of the early nineties, alarmed by what they misperceived in the men's work as reversion to primitive forms of male savagery, joined with the Right to discredit and attack the mythopoetic work, shaming many men into prematurely closing off this avenue of exploration. That feminists should thus inadvertently aid the cause of a cynical patriarchal status quo is one of the more tragic ironies of our time.

A widely held and damaging false assumption about the men's mythopoetic soul work was that it proposed to challenge or repeal the gains women had worked hard to achieve through decades of political effort. But the mythopoetic work, as developed by Bly, Michael Meade, James Hillman, Robert Moore, and others, had no such goal. In 2000, the *Minneapolis Star Tribune* asked Bly which activities he'd recommend for a national day for boys equivalent to Take Our Daughters to Work Day. Bly's reply was utterly consistent with the vision of men's work he'd promoted fifteen years earlier. He suggested that fathers take their sons to the library and show them books they love. Acknowledging that women have often been excluded from the work world, Bly remarked, "I think it's just as likely now that men will be shut out of the inward world, the literature world."

In grasping the changes of the past decade, it may help to remember that 1992 was the year Bill Clinton and Al Gore were elected to their first term in the White House. In his best-selling volume of environmental advocacy, *Earth in the Balance,* Gore relates the incident of his son's near-fatal accident in 1969. Quoting *Iron John,* Gore writes that he went down "into the ashes," a reference to the protagonist's moment of truth in the Grimm Brothers fairy tale. I wasn't the only one to notice this new note being sounded by the Democratic candidates. In the *New York Times* for 27 July 1992, Maureen Dowd wrote breezily of the Bush-Quayle versus Clinton-Gore ticket that "the fall offers a fascinating contest between John Wayne Republicans, never crying or even blinking, and Iron John Democrats, seeking the inner man and the missing father."

The contrast Maureen Dowd remarked in 1992 recurred even more starkly in the campaign for the 2000 election, an opposition personified by the psychologically and environmentally astute Gore and the crude Bush Jr., throwback to the clueless macho male of the 1950s.

I mourn the plight in which our young men now find themselves, often lacking serious role models and mentors, and virtually abandoned by parents to an economic order in which, as the Beat writer William Burroughs accurately observed, "Instead of selling the product to the customer, we sell the customer to the product." Shame and anger are written in the sullen expressions and slack postures of today's youth, a vacuum of self-esteem into which too easily rushes the warmongers' mystique of redeeming violence. How much better it might be for all our children now if the mythopoetic work had taken a firmer hold in the soil of the malls and the new entertainment/warfare state.

The Myth of Phaethon

The Phaethon myth has much to say about the world of the twenty-first century and how it came to be the way it is. Here is a glance at Ovid's version:

Phaethon is the young son of the solar god Phoebus Apollo and Oceanus's daughter Clymene. Raised by his mother, Phaethon feels unsure whether Phoebus actually is his father, and early one morning appears in Phoebus's temple to establish that lineage. To prove that he is indeed Phaethon's father, Phoebus vows to grant any favor Phaethon requests. The boy asks to be allowed to drive the chariot of the sun that Phoebus himself guides across the sky every dawn, borne aloft by a team of winged horses.

Phoebus instantly regrets his pledge, yet feels honor-bound to fulfill it. The horses are formidably powerful, he warns, and often even he himself can barely contain them, trembling at the fearful task of keeping the chariot on course so high above the earth. Moreover, the driver must not pass too close to all those fearsome beasts of the Zodiac—the Bull, the Scorpion, the Dragon, and so on—for fear of being drawn down into their clutches. Phoebus urges, "I do not want to give you / The gift of death, there is time to change your prayer. . . . / It will be given, whatever you choose. I swore it. / But choose more wisely!" (All quotes in this essay are from the elegant Rolfe Humphries translation of *Metamorphoses*, Indiana University Press.)

The son is unconvinced. Suppressing dread, Phoebus instructs Phaethon in the subtle celestial navigation he must instantly master, then hands over the reins, for it is dawn, time to drive the exquisitely wrought chariot of the sun. In parting, Phoebus advises Phaethon to "go easy on the whip, hard on the reins": the horses need no encouragement but only a firm hand. Phoebus specifies the path Phaethon must follow midway between earth and sky: "Sky and earth both need / Equal degrees of heat: too low, you burn / The one, too high the other."

But youth can seldom hear the adult wisdom of moderation. Just as Phoebus has feared, the overconfident Phaethon almost immediately loses control of the horses, who, feeling a lighter touch on the reins, go their own way. At this point, Ovid says of Phaethon, "the darkness came / Into his eyes from too much light." Paralyzed by fear, Phaethon drops the reins. The unguided chariot careens chaotically between Heaven and earth, immolating both:

> . . . The scorched clouds smoke. The mountains
> Of earth catch fire, the prairies crack, the rivers
> Dry up, the meadows are white-hot, the trees,
> The leaves, burn to a crisp, the crops are tinder.
>   . . . The great cities
> Perish, and their great walls; and nations perish
> With all their people: everything is ashes.

Thus an apocalyptic scene of global destruction results from the fulfillment of a boy's grandiose, innocent wish. It's the Waste Land that has haunted the Grail poet Wolfram von Eschenbach and T. S. Eliot, as well as all of us who now live in an era haunted by the menaces of nuclear weapons and global warming.

>   . . . Even the ocean
> Shrinks to a plain of sand; the hidden mountains

Emerge to join the Cyclades; the dolphins
Dare leap and curve in the high air no longer;
The fish dive deep, and the dead seals are floating,
White-bellied on the surface. The story has it
That Nereus and Doris and their daughters
Found even their deep-sea caverns hot and stifling.
Neptune, with scowling countenance, dared lift
His arms, three times, above the waves; three times
He could not bear the fiery air.

Finally we hear the voice of Earth herself grieving the holocaust. Her heartbreaking plea to Jove, as rendered by the Roman poet in the first century of our era, reproaches us powerfully in our dawning millennium:

"O greatest of the gods, if this is pleasing
And I deserve it, why hold back the lightning?
If I must die by fire, then let me perish
By fire you send, and lighten the destruction
Because you are its author. I can hardly"—
The smoke was suffocating—"open my lips to speak;
Look at my hair, burned crisp; look at the ashes
In eyes and face! Is this what I am given
For being fruitful, dutiful? for bearing
The wounds of harrow and plowshare, year on year?
Is this my due reward for giving fodder
To flocks and herds and corn to men, and incense
For the gods' altars? Maybe I deserve it,
But what about the ocean, and your brother?
Neptune's allotted waters ebb and vanish,
Farther and farther from Heaven. Well; never mind him,
Never mind me, but have a little pity
For your own skies. Look! On both sides the poles
Are smoking. If that fire corrupts the heavens
Your palaces will topple. Even Atlas
Strains and can hardly bear his white-hot burden.
If sea and land and sky are lost, we are hurled
Into the ancient chaos. Save us, father;
Preserve this residue; take thought, take counsel
For the sum of things."

Rendered speechless by the heat, Earth lapses into silence. But her words have moved Jove to intervene on her behalf. His lightning bolts shatter the chariot from beneath the luckless Phaethon, who, "His ruddy hair on fire, falls streaming down / The long trail of the air." The Western Naiads bury Phaethon's remains. His mother laments bitterly, and Phoebus mourns so inconsolably that an entire day passes without sun. Phaethon's sisters become so distraught that, while continuing to weep, they turn into trees; their tears become amber and fall into the river, to eventually adorn Roman brides. Meanwhile, Jove attends to restoring the devastated earth.

Senex, Puer, and Earth

Perhaps most obviously, Phaethon's catastrophic fall can be read as a cautionary tale in the vein of the myth of Icarus (also recounted by Ovid). In that story, Daedalus, the artist and inventor recruited to build the Labyrinth for King Minos of Crete, longs to return to his homeland, but Minos prevents his passage by ship. Preparing to escape by air with his son, Icarus, on wings fashioned from feathers and wax, he echoes Phoebus, advising the boy to fly neither too low nor too high but to adhere to a safe "middle course." All goes well until, exhilarated by youthful high spirits and the sheer joy of flight, Icarus soars too near the sun. In Breughel's haunting interpretation, Icarus's fatal plunge into the sea is witnessed with mild wonder by fishermen and ploughmen going about their ordinary business. In both stories, the son dies as a consequence of misuse of a flying device associated with the father.

One of the conventions of the Jungian and archetypal analysis applied by von Franz and Bly is to view every character in the story as representing an element of the psyche. Viewed in this way, the Icarus and Phaethon stories point toward the presence in each person of two contrasting, yet complementary male-toned energies known in the Jungian vocabulary as the Senex and the Puer. James Hillman has written of them brilliantly in *Puer Papers* (1979). The Senex is the seasoned old man who possesses knowledge and life experience; yet he is in some way dried up, stringy, sardonic, slow-pulsed, cold-blooded. The Puer is the youth whom popular culture admires as much as the rock star or athlete; he is brash, young, ambitious, swift, hot-blooded, juicy. Both have their peculiar strengths and weaknesses. Though life in the Senex has ebbed, he holds the key to life's wisdom-granary. The Puer has little lived experience but an attractively abundant appetite for acquiring it. The Senex is about limits, while the Puer is about defying limits.

The story of Phaethon suggests that, on an intrapsychic level, some callow, youthful part of ourselves wants to unseat or usurp the power of the part of us that is old, tempered, and wise. The Senex in the psyche is the one who has learned, through a long lifetime, how to live on and with the earth, who knows and respects the force of gravity as the untested

Puer does not. If the Senex energy is that of gravity that holds us to earth and the ways of nature, the Puer energy seeks to outrace gravity, to leave earth (and by implication, embodiment itself), to *become* spirit. For the Senex, recognition of limitations also means a conscious acceptance of death as a part of life. Because the Puer's consciousness does not yet include that recognition of limits and the role and presence of death in life, he is dangerously vulnerable in his flightiness.

We enact the Puer as a culture in our manic pursuit of speed and mobility (as concretized in our jets and automobiles) at any cost, including the poisoning of earth, air, and water in our demand for an unlimited supply of fossil fuels. In our willingness to despoil the environment upon which all living creatures depend, we embody the Phaethon myth rather than the Icarus myth, in which Icarus dies but the world itself is unharmed.

The Senex in the psyche also contains and perpetuates the ancestral and indigenous knowledge of the human race, which is closely allied with nature. By contrast, the Puer's infatuation with novelty, speed, and risk tends to lift us away from the earth and its necessity, which human beings have often grudgingly embraced in agreeing to live on earth. In this respect, the Puer is radically individualistic, while the Senex embodies the basic conservatism of community.

The Puer's urge for escape from gravitic restraints is by no means negative in itself, even though it proves fatal to both Phaethon and Icarus. In fact, the Puer is one of our greatest internal generators of creativity and innovation on both individual and cultural levels. His impulse toward flight is explicitly related to the divine or solar intelligence within the self, so magnetically attractive to the spiritual Puer. The stories of Phaethon and Icarus should not be interpreted as against the impulsively beautiful young-man energy of the Puer but rather as instructive of the intricate balance we all need in negotiating and integrating these powerful energies within ourselves. Without the Puer's counterbalancing injection of novelty and risk, the Senex alone becomes sterile, even antilife.

For our present age, the Phaethon story may be most useful for its unforgettable and horrifying images of a planet consumed by ecological disaster, which is our own. In 2000, the Ninth International Coral Reef Symposium in Bali reported that an estimated quarter of the world's coral reefs have died due to an increase in ocean temperatures, with the remainder to follow, if we do not act on their behalf, within about twenty years. Polar bear populations are diminishing as a result of the shrinkage of ice floes that function as their hunting grounds. The list of endangered species continues to grow at a shocking rate, with extinctions worldwide averaging one per day, and threatened rain forests continuing to fall to the developers' bulldozers. Recently, the Pacific Island nation of Tuvalu announced plans to evacuate its entire population to New Zealand because of rising sea levels. The eleven thousand people of Tuvalu, we may sadly predict, will not be the last human victims of our collective uncontrolled chariot ride.

In our story, it is the prayer of Earth herself that persuades Jove to end Phaethon's nightmarish spree. This archaic vision of earth as a living being echoes down to our own time in James Lovelock's Gaia Hypothesis, which proposes in scientific terms that earth itself is a vast living organism. Freakishly severe weather and geological upheaval worldwide should serve notice that Earth's prayers are not to be taken lightly. In today's earthquakes, hurricanes, and tornadoes, we hear Earth's prayer to Jove: " . . . Save us, father; / Preserve this residue; take thought, take counsel / For the sum of things." "The sum of things" is of course the province and concern of ecological consciousness, of which we habitually lose sight in our pursuit of unlimited consumption, bringing down upon ourselves shame as men and as human beings.

Coda: Protection

Why, we might ask, is Phoebus unable to prevent Phaethon from committing the horrible mistake we can easily see coming? Does the answer lie at least partly in the limited nature of rationality itself, which Phoebus Apollo classically embodies? We can recognize something of Phoebus's inability to protect both his son and the earth in our own apparent inability to defend our children and the environment against the inhuman extremes of the technological mentality. Where earlier ages justified their actions by divine authority, we justify our actions by the authority of the rational intellect. In terms of survival as a species dependent upon an ecosystem, our faith in rationality has led to practical *irrationality*. Is it "rational," for example, to delay responding to the threat of catastrophic global climate change until it may be too late for our response to be effective? Is it "rational" to scrap hard-won civil liberties under threat of a terrorist attack statistically many times less likely to kill us than a drive on the freeway? Is it "rational" to continue unreflectively consuming when we know that earth's resources are finite and come at a mounting expense of human misery and damage to our environment and ourselves?

In these matters, I suspect that we all feel some profound shame, if only on an unconscious level. As a society, we are like Phoebus in the story when we hand our grade-school children over to computers without knowing how that will affect their health, intelligence, and spiritual well-being. The unexamined, relentless drive to "wire" our schools is one among many possible examples of Burroughs's "selling the customer to the product." In the name of a materially defined progress, we have sold our children to rapacious consumerism at a very low price.

In the past quarter-century, men have faced a loss or weakening of their traditional roles in a void of positive new roles to replace them. In a time in which girls and young women are encouraged to view the world as a place of "girl power" and unlimited opportunity for self-realization, boys and young men too often see a contracting horizon of diminished expectations. The rage elicited in young males by this hopelessness and disappointment is easily harnessed and channeled by the cynical military establishment as aggression toward perceived enemies.

I still find value in the traditional masculine desire to protect one's family, one's community, one's planet. With the New Warriors, I believe that there is great untapped potential for a "positive" new masculinity in standing up to the military-industrial complex in defense of the earth. Older men who take on the burden of this defense can provide our growing number of ungrounded, dispirited boys with a true sense of purpose and self-worth in helping to create a less violent and less violated world.

As a child of the 1960s, I've often wondered why the ecological consciousness of my generation hasn't translated into a more environmentally activist political milieu, more willing to meet the pressing challenges of the twenty-first century. When my generation should have been coming into its own, the Reagan administration initiated a series of economic assaults on middle- and lower-income people that continues to this day; distracted by consumer bounty and constant worry over loss of jobs and homes, we have lost sight of the larger picture. In this climate of overconsumption coupled with radical insecurity, it's no wonder we scrabble to make ends meet while leaving the Big Picture in the hands of compromised politicians.

Finally, we must ask ourselves a hard question: What is the result when men abandon responsibility for stewardship of the earth? We glimpse one answer to that question in events such as the 2002 midterm elections, in which issues of "homeland security" and fear of terrorism won the anti-environmentalist, prowar Bush administration a degree of power unrivaled in recent history. Bush's campaigning for Republicans around the country succeeded precisely because Americans feel physically and psychologically vulnerable after the September 11th attacks, and the Right Wing promises to protect them. Though the ironies of this posture are apparent when one considers how much more dangerous the world seems now, and how much more fearful we are now than when Bush assumed office, the fact seems clear: Lacking the courage to elect leaders who, in the true spirit of democracy, represent us, we instead elect the biggest "neighborhood bully" around to protect us. This gives us a temporary security fix, but something gnaws at us internally because we know that earth, our life-support system, is dying, and we are doing nothing to save it.

Abdicating representative democracy in exchange for a protection that no one in this explosive and interconnected world can actually guarantee us, we deliver life and the planet over to the destroyers. In our own lack of courage to take hold of the reins, we watch Phaethon careen out of control in his fatal flight, precipitating fiery chaos. Only by reclaiming each person's individual responsibility for being guardians of earth and a hopeful future for the children of all living species, can we manifest the collective courage to restore real democracy and thereby shed the unspoken shame that is one of our most spiritually corrosive secrets.

Works Cited

Bly, Robert. *Iron John: A Book about Men.* Reading, Mass.: Addison-Wesley, 1990.

Hillman, James. "Senex and Puer: An Aspect of the Historical and Psychological Present." In *Puer Papers.* Dallas: Spring Publications, 1979.

Meade, Michael. *Men and the Water of Life: Initiation and the Tempering of Men.* New York: HarperSanFrancisco, 1993.

Ovid. *Metamorphoses.* Translated by Rolfe Humphries. Bloomington: Indiana University Press, 1955.

Sardello, Robert. *Freeing the Soul from Fear.* New York: Riverhead Books, 1999.

Taking care
Toward an Ecomasculinist Literary Criticism?

SCOTT SLOVIC

I'm a scholar and teacher who, in general, pays little conscious attention to gender as a crucial determinant of literary expression. I tend not to think of how gender contributes to my own thinking about language, my social interactions, and my relationship to the natural world. Sure, I'll freely admit it's possible that much of what I do is guided by my society's gender norms or by biological coding, but the issue of gender is seldom an explicit, conscious part of my work or my daily life.

I have few claims to authority as a scholar practicing gendered approaches to literature or gendered approaches to anything at all. More than a decade ago, when I was studying Annie Dillard's work for my dissertation, research that became part of my first book, *Seeking Awareness in American Nature Writing*, I was not at all interested in the gender dimension of her writing, but I found myself reading some things that have stuck with me all these years, always confounding and troubling me, making me worry that perhaps gender was a more significant issue than I was acknowledging. I'm thinking specifically of Dillard's account, presented in her essay "To Fashion a Text," written in the mid-1980s, of her gender contortions in drafting her famous early book, *Pilgrim at Tinker Creek*. According to her account in "To Fashion a Text," Dillard began toying with the manuscript that eventually became *Pilgrim* while hiking in Acadia National Park in Maine, and in these early musings and mental notes, the author imagined adopting the voice of a male narrator in her work. She wrote the opening chapter in the first person, as a man, before giving up the plan. What always confused me about Dillard's initial plan was not only the notion that one could write a book about metaphysics and natural history from an explicitly gendered perspective, but that one could possibly change the gender of one's literary voice like changing clothes. I have wondered what the early manuscripts of Dillard's book might have looked like, especially what it meant for her to write "in the first person, as a man" (57). Nonetheless, when I pursued my analysis of the motif of consciousness in Dillard's writings, I gave the issue of gender little attention.

However, the idea did continue to trouble me. What *was* a "male narrative voice"? *Was* there such a thing? Surely gender was potentially part of the subject one could choose to write about—a writer could choose to foreground gender matters or deemphasize them. I chose, as a critic, to deemphasize gender and to work with authors who, it seemed to me at the time, did not make gender the focal point of their work. But, as most readers know, as soon as you start looking for something in literature, you realize it's there. I honestly wasn't looking for gendered themes when I examined works by Edward Abbey, Wendell Berry, and Barry Lopez for the consciousness motif. But now, when I put on the lens of a gender critic, I see masculinity—or the playful scrutiny and mockery of masculinity—almost everywhere in their work (yes, especially in Abbey's work). Still, this "gender lens" is a recent thing for me, and not a comfortable perspective—I doubt it will ever be so with me.

At the time when I was working on Dillard, back in the late 1980s, I also read a memorable essay on Dillard by Suzanne Clark. The essay, titled "Annie Dillard: The Woman in Nature and the Subject of Nonfiction," explained that even though Dillard made such an issue of distorting and contorting the gender of her narrative voice, this very effort of self-abnegation, of removing herself from her biologically feminine voice, was something a female writer would opt to do, not a male writer. Clark makes quite an elegant, compelling argument that a writer is forever, intrinsically a product of his or her gender—writing from the experience, from the perspective, of one's gender is inevitable. As Clark puts it:

> When we read Annie Dillard, we don't know who is writing. There is a silence in the place where there might be an image of the social self—of personality, character, or ego. There is little mention of her sex or her work or her success, or even—until later works—of the color of her hair. The voice of realistic fiction and the objective voice of fact are commonly anonymous and faceless in the service of giving us the illusion of nonbias. But this lack of self in Dillard is not a lack of subjective bias. Indeed, her prose style enacts a stylistics of subjective bias, writing out the very gesture of perception as a kind of poetics. This "poetics" makes perception and self-identity figurative. At the very place of lack, there appears a figure of self-consciousness. Women writers commonly find themselves caught in a discourse which excludes the woman as author, even if the author is a woman....

The difficulties of calling Dillard's voice a woman's voice have not, in truth, seemed especially visible to other readers. In spite of the fact that Dillard's narrator does not call attention to her gender, and in spite of the striking absence of female detail, Dillard's writing has already drawn attention as woman's writing. But is this because we know Dillard to be a woman writer, or because she writes like a woman in some ways we can specify? (107–8)

Baffled by the notion that even writers who take pains not to make themselves—or their gender or ethnicity—the focus of their work might be "found out" by critics keen to discern the inevitable nuances of self or sex in literary prose, I figured I'd just stay away from this issue. This scholarly approach seemed ingenious and yet somehow unfair to the purposes of the work at hand, the goals of the author, insofar as these things mattered. I managed to put this argument aside and focus on other things in my work, other aspects of environmental literature, not understanding how the logic of Clark's claim applied not only to Annie Dillard or to the many other male and female authors I've studied and taught over the years but even to myself. Perhaps it's true, too, that I can't actually do anything without being of a particular gender—perhaps even the presumption that I can (or could) look at particular aspects of literature in general, or environmental literature more specifically, without concern for gender is a *male* perspective. In other words, because I am a heterosexual male critic, some might argue that I have the "privilege" of not being concerned with gender, while women critics, or other kinds of scholars whose gender is somehow more vexed or complicated, such as gay and lesbian scholars, cannot help but think consciously about gender and sexual preference in their work or as they read the work of others.

I feel I'm quickly getting in over my depth here, but I am simply trying to trace my thinking as I work my way to the main points I'd like to make in this essay. I'm sure plenty of scholars have already figured out a vocabulary for discussing these topics. Perhaps the naiveté of my own musings will be compensated for by the freshness of mostly jargonless speech. I guess what I've been trying to say is that I've been haunted in a way, for the past decade and a half, by Dillard's remarks about her contortions of gender in crafting the narrator of *Pilgrim at Tinker Creek* and by Clark's commentary on Dillard's contortions—I have wondered, at the barely conscious level, whether my approach to literature has been a *delusion* of gender blindness rather than the detached, neutral perspective I've always imagined it to be.

June 2001, Flagstaff, Arizona. One afternoon during the fourth biennial conference of the Association for the Study of Literature and Environment, two curious things happened within an hour of each other. First, there was a plenary session devoted to the work of Annette

Kolodny, the pioneering feminist scholar of American literature whose major books include *The Lay of the Land: Metaphor as Experience and History in American Life and Letters* (1975) and *The Land Before Her: Fantasy and Experience of the American Frontiers, 1630–1860* (1984); several of Kolodny's current and former students spoke about her importance to the field of ecocriticism, and then Kolodny offered her own remarks about gender, literature, and nature. During the plenary session, Kolodny asserted that many of the world's environmental problems can be attributed to "the male lust for El Dorado," perhaps intending this phrase as tongue-in-cheek hyperbole or perhaps not. Immediately following the Kolodny session, the audience of several hundred scholars broke up for an assortment of concurrent panels, one of which was the session called "Men and Nature: Perspectives on Masculinity and the More-Than-Human World," which Mark Allister and I had organized. Mark and I joined nature writers Ken Lamberton and Gary Paul Nabhan to see what might be done to rehabilitate the reputation of half of the human species. Admittedly, with George W. Bush and Dick Cheney running the United States government these days, this is an uphill struggle. Gary showed up for our panel with a burlap bag slung over his shoulder—before sitting down at the speakers' table, he removed a tire and a bottle of beer from the bag, kicked the tire (a small tire from a wheelbarrow) and opened the Negra Modelo. Then the men were ready to speak.

Mark and I had first started communicating about the idea of the panel on men and nature when he contacted me to respond to my essay "'Be Prepared for the Worst': Love, Anticipated Loss, and Environmental Valuation," concerning the death of my infant son several years earlier. The idea of the ASLE panel lit a spark in my imagination. "Men and Nature"—what a weird idea! What an obvious idea! I was quite sure there had never yet been such a panel at an ASLE Conference, and I couldn't recall seeing such a panel at any of the many conferences I had attended or hosted over the years—Western Literature, Environment and Community, American Studies Association, MLA. To this day, I can't even recall seeing panels on "men and literature" at any of these meetings, although I can think of plenty of sessions devoted to women's writing, sexuality in general, or sexual preference (normally, gay and lesbian approaches or perspectives). It's as if the issue of heterosexual, masculine perspectives on anything—literary expression, culture in general, the world beyond human beings—either does not warrant discussion because masculine perspectives are assumed to be obvious and pervasive, taken for granted somehow, or because masculinity is so controversial—so politically incorrect, self-justifying, embarrassingly offensive—that nobody wants to talk about it, at least not in mixed company.

It appears to be taboo to use the phrase "men and nature" in anything other than a derogatory way. This seems irresponsible to me, blindly dismissive. Sure, there has been plenty of male "lust for El Dorado" throughout history and in every part of the world—and plenty of female lust for wealth and plunder as well. However, in academic circles it has become universally fashionable to complain about male destructiveness. Human ecologist Garrett Hardin has written extensively about population growth as a taboo subject, suggesting in *The Ostrich Factor* that "Taboo discourages us from taking a total view of the effects of size on the well-being of human populations. For this willful blindness, society ultimately pays a price" (15). Likewise, I would argue, the scapegoating of men, the habitual identification of stereotypical male attitudes and behaviors with social and environmental destruction, has become a kind of mindless reflex in feminist circles and more broadly in socially progressive circles. And talking openly about such scapegoating has become taboo in liberal communities, just as population is a taboo topic among many conservatives.

"Men," "male," "masculine"—these words and the categories of being represented by them have become "otherised" by the prevailing discourses of feminism and environmentalism. In her recent book *Environmental Culture: The Ecological Crisis of Reason,* Val Plumwood writes at length in a chapter called "The Blindspots of Centrism and Human Self-Enclosure" about the dangers of "otherising" entire categories of people, building upon Edward Said's famous critique of eurocentric "Orientalism." Plumwood argues that "Radical exclusion marks the Otherised group out as both inferior and radically separate. The woman is set apart as having a different nature, is seen as part of a different, lower order of being" (101–2). One of the unspoken blind spots of ecofeminist discourse is its tacit inversion of traditional, European, male-centered hierarchies of value. In other words, although scholars have been reluctant to admit this, there is an implicit argument in ecofeminism that women are morally superior to men by virtue of their historical subjugation in certain cultures. One might argue, by extension, that the route to absolution for male environmentalists is to identify themselves as ecofeminists, buy into the male-female dualism by attributing past and current environmental and social ills to patriarchal domination, and through the process of critiquing their own biological gender metamorphose their sense of self into a transcendent, morally superior meta-gender.

In the introduction to their 1998 collection *Ecofeminist Literary Criticism: Theory, Interpretation, Criticism,* Greta Gaard and Patrick D. Murphy begin, reasonably, by acknowledging that "Ecofeminism is not a single master theory and its practitioners have different articulations of their social practice" (2). But in the process of clarifying what is meant by the term "ecofeminism," they proceed to quote four key points made by Ynestra King in her foundational 1990 article "Healing the Wounds: Feminism, Ecology, and the Nature/Culture Dualism," suggesting that these "basic principles ... have been quoted extensively and generally embraced as a sound orientation" (3). The first of these principles argues that "Western industrial civilization" has traditionally viewed women as closer to nature than men and this has "reinforced" the subjugation of women. "Therefore," writes King, "ecofeminists take on the life-struggles of all nature as our own" (*Gaard and Murphy* 3). Although she delicately avoids saying "therefore *women* take on the life-struggles," leaving open the possibility of male ecofeminists, this first principle of ecofeminism buys into the very dualism (the opposition of men and women) that is attributed to patriarchalism, pivoting off of this abhorrent dualism and reinforcing it at the same time. King's first principle of ecofeminism seems directly in conflict with the second principle, which begins, "Life on earth is an interconnected web, not a hierarchy" (3), making this statement appear disingenuous. The third and fourth principles advocate diversity and the rethinking of human relationships to nature "according to feminist and ecological principles" (4). Principles two, three, and four would be of wide appeal to people of all cultures, genders, and socioeconomic strata, except for the fact that, as articulated by King, they follow the opening (and perhaps inevitable) critique of patriarchy. King is here following Karen J. Warren, who in her 1987 essay "Feminism and Ecology: Making Connections" offered the following statement:

> A *patriarchal conceptual framework* is one which takes traditionally male-identified beliefs, values, attitudes, and assumptions as the only, or the standard, or superior ones; it gives higher status or prestige to what has been traditionally identified as "male" than to what has been traditionally identified as "female." A patriarchal conceptual framework is characterized by *value-hierarchical thinking.* (Quoted in Hay 73)

Although it is perhaps inevitable that a social movement should root itself in some form of critique, the way in which this critique is voiced has everything to do with how widely the views of the movement are embraced. It is all too common that visions of social reform are expressed in language so angry and self-righteous that potential supporters are put off, scared away, or otherwise disenfranchised. Discussions of nature and gender, at least since the mid-1970s, have focused almost exclusively on critiquing what Warren calls "traditionally male-identified beliefs, values, attitudes, and assumptions" and celebrating "what has traditionally been identified as 'female.'" Scholars, artists, and activists have been left with two viable, politically correct choices: identify with the ecofeminist critique of patriarchy/masculinity or keep silent with regard to issues of gender.

I believe the time has come, at least within the community of ecocritics and environmental writers, to pay attention to the many examples of positive, healthy male attitudes toward other humans and toward the natural world available in contemporary environmental writing. I am reluctant to label the attitudes expressed in my examples below as specifically "male" because I believe they are, in fact, universally accessible and acceptable views of the world. For the purpose of correcting what I take to be the hypercritical attitude toward men in gender-conscious literary scholarship during the past quarter-century, I will mention several books and essays from the past decade that show men in caretaking roles, particularly as fathers or as children looking after aging parents, but also as environmentally conscious individuals demonstrating a caring, careful view of the more-than-human world. Although some of these texts emphasize the importance of protecting, nurturing, and guiding other human beings, it is easy to extrapolate from the human relationships depicted in these works and extend the narrators' attitudes to their relationships with things in nature. Perhaps this seems like a simpleminded, anecdotal rebuttal to the male-lust-for-El-Dorado perspective. But the several samples discussed below are merely a few of the many examples of constructive male voices, constructive male stories, available in environmental literature. The upshot, I would say, is that all of us—scholars, writers, teachers, students, activists—ought to "take care" to avoid dismissive, essentialist statements when we seek to pinpoint environmental and social culprits (and heroes). For every person who uses the verbal shorthand "men" (or "male") to describe the source of all ecological degradation, there's another who is inclined to attribute the world's problems to "those Americans" or to "Westerners" or "whites," often as a way of nullifying the speaker's own complicity. This brief examination of male caretaking in contemporary environmental literature aims simply to halt a reflexive antagonism toward men, or the "idea of men."

We can take critique only so far. I believe it is necessary to imagine positive roles, positive voices, for all people to have, regardless of gender, as we continue to struggle, in fits and starts, toward a more enlightened place for our species on this planet.

One mode of undertaking this process—still in the vein of corrective to the excesses of ecofeminism—would involve the establishment of an "ecomasculinist literary criticism" that would seek to identify, in literary texts, socially and ecologically responsible behavioral and linguistic models for men. If one wanted to take a hard-edged feminist view of the examples I'm going to give, I'm sure it would be possible to find things to disparage and discard. But that's not my approach to the world—I would rather try salvaging good things and bringing as many people as possible into the discussion. I could easily name several dozen interesting and constructive examples of what I take to be positive contemporary male voices. A thumbnail survey would include the following examples: Gary Paul Nabhan and Stephen Trimble's braided collection of essays, *The Geography of Childhood: Why Children Need Wild Places* (1994), presenting their own childhood experiences and reflecting upon their efforts to expose their own (at that time) young children to wild nature; Harold P. Simonson's book of essays *Going Where I Have to Go: Essays from Within* (1996), particularly such pieces as "Risking It" and "Appointment on Everest," which examine why people are compelled to climb mountains from the perspective of a father of mountaineering sons; John Elder's *Reading the Mountains of Home* (1998) and Scott Russell Sanders's *Hunting for Hope: A Father's Journeys* (1998) show that the love of family and love of place are versions of the same emotion; David James Duncan's multivalent collection of prose meditations—angry, despairing, celebratory—*My Story As Told by Water* (2001); and specific recent essays such as Rick Bass's "The Farm" (2001) and Paul Lindholdt's "The Spray and the Slamming Sea" (2002), which focus on the joys and terrors of being a parent. I find myself thinking, too, of Barry Lopez's extraordinary recent essay "A Scary Abundance of Water: Growing Up with the San Fernando Valley" (2002), which reveals the nuances of the author's childhood pain and joy, links these pains and joys to the experience of the society more generally, and asserts:

> The peculiar task of many American writers today—though, again, only as I see it—is to address what lies beyond racism, class structure and violence in American life by first recognizing these failings as real, and then by helping with the invention of what will work in such circumstances to ensure each life endures less cruelty, that each life is less painful. (31)

Each of these books and articles by contemporary male authors defies the standard criticisms of masculinity. Each, in its own way, is a demonstration of "taking care," showing sensitive engagement with other people and with other natural phenomena, while also expressing, in many cases, eloquent social analysis and critique. At the conclusion of her well-known study *Feminism and the Mastery of Nature,* Val Plumwood calls for "remaking" the "reason/nature story [which] has been the master story of western culture" (195–96). The authors I have just mentioned are engaged in precisely this process of undoing the objectifying, colonizing dualism between reason and nature, sensitive to the implications of their new stories.

A full articulation of "ecomasculinist literary criticism" would entail thorough examination of numerous texts in an effort to identify the special male virtues of these texts that make them exemplary in their transcendence of problematic traditional tendencies of male behavior and male discourse vis-à-vis society and the planet. My purpose here is to make a preliminary gesture toward rehabilitating the reputation of male writers about nature, and I have selected two texts to discuss in somewhat more detail, because the very titles of these books say basically what I want to say. The first is John Daniel's 1996 book *Looking After: A Son's Memoir*; the second, William Kittredge's 1999 Credo essay *Taking Care: Thoughts on Storytelling and Belief.*

Daniel's book is an intertwined memoir of his own move from the eastern United States of his childhood to his adult life in Oregon and California and an account of his mother's final years of life, after she came west and moved in with John and his wife, Marilyn, who were then living in Portland, Oregon (they've since moved to the small town of Elmira, west of Eugene). Although Daniel is widely regarded as a major contemporary American environmental writer, one would find it difficult to argue that this particular book is "nature writing" in any traditional sense of the term. Yes, there are lovely, lyrical descriptions of the eastern Oregon high desert and of trees and birds viewed through the windows of the Daniel family home in Portland, and there are moving descriptions of the human body and the fickle and frightening human mind—surely parts of "nature" as well. But for me, the deepest "environmental" dimension of this work is in the representation of John's relationship with his mother, Zilla Daniel, a strikingly independent woman, even into her seventies. The book is full of tenderness and remorse, and full of darker emotions such as impatience and frustration. Zilla ended up living with John and Marilyn for the last five or six years of her life. During this time she fell deeper and deeper into the helplessness of Alzheimer's, and her son and daughter-in-law became increasingly responsible for her every need. But what emerges in the narrative, at least in my reading of it, is the awesome, positive potential for people—men and women—to "look after" what they love, to care about another person's life enough to remember it and prepare the kind of written document we call "memoir" and to care enough to tend to the most banal and awkward requirements of daily life.

One of my favorite passages in the book is an odd little segment about the author's experience helping his mother take a shower. Here is the final part of that story:

> I helped my mother down into a straight-backed chair and left her in the bathroom with towels, clean underwear, and a little space heater to keep her warm. She took her time, as with everything. Often it was half an hour before she emerged in her dressing gown, her hair beginning to fluff, her face smiling. No matter how hard she might have resisted the idea, a bath or shower always seemed to renew her. Soap or no soap, the old woman came forth cleaner of spirit.
>
> "She was pure as the driven snow," she usually quoted, gaily, then a pause: "But she drifted."
>
> I guess I came out of the bathroom cleaner of spirit myself. Soap or no soap, whatever the tenor of our conversation, I appreciate now what a privilege it was to help my mother with her shower. I wish I'd seen it more clearly at the time. We don't get to choose our privileges, and the ones that come to us aren't always the ones we would choose, and each of them is as much a burden as joy. But they do come, and it's important to know them for what they are. (75–76)

Daniel wrote this book as a form of tribute to his mother and as a way of thinking through his own quest for a comfortable understanding of personal identity—it's not always a pretty self-portrait, but it's not maudlin or pettily self-deprecating, either. It feels candid, and the abiding emotion is tenderness—and this from a towering ex-logger and rock climber. Despite the parts of the book in which the author criticizes his own weaknesses or excesses, this is a narrative that ultimately reveals the redemptive process of "looking after." Certainly one could argue that this redemption transcends the social and environmental role of any one gender—but since the function of "ecomasculinist literary criticism," as I have identified it, is to rehabilitate the role of men in the world, it may be enough to argue that John Daniel's *Looking After* depicts a man trying to have a supportive, caretaking role within his family and in the world more generally. Daniel extends the process of self-analysis in his recent publication, *Winter Creek: One Writer's Natural History*.

Taking Care is the fourth volume in a series of five self-admonishing meditations spun out by William Kittredge from the mid-1980s up to the present. Self-admonishment, confession, plea—it's hard to know just what to call *Owning It All*; *Hole in the Sky*; *Who Owns the West?*; or the more recent volumes, *Taking Care* and *The Nature of Generosity*. I think one positive role for men to play in the world is the looking-after role John Daniel mentions in the title of his book—the role of protecting, remembering, and guiding (even the guiding of one's parents through the end-of-life joys and throes). Another positive role men can play in the world is that of "taking care." Kittredge's book titles often have double entendres, or sometimes several meanings. The most obvious implication of the title *Taking Care* is the author's belief in the importance of living carefully, gently, and nondestructively. Of course, this concern for "taking care" is the result of having lived a life that, for many years, involved very little care or respect for land, animals, or even, some would argue, other people. Kittredge's earlier books in the confessional series are excoriations of his family, and implicitly himself, for partaking in a Western mythology that was, essentially, an excuse for abusing and exploiting natural resources and human beings. As Kittredge puts it:

> In springtime I had to admit that the waterbirds were not going north in huge flocks anymore. Peat soils were going saline and blowing away. I had to admit that my work was one of the reasons these things were happening. I had to embrace my responsibility for the wearing out. . . .
>
> But I couldn't imagine doing my work according to any other model. Born to the ruling class, it was my responsibility to make the world go; I had no other work. I was fated, or so I thought, to participate in the diminishment of things I cherished, and that idea made me crazy. (26)

After watching the family's agricultural practices in southeastern Oregon destroy the land's fertility, and after a series of traumatic personal relationships, the author began to come to his senses and develop the ethic of care and respect articulated in the title of the 1999 book. If one wanted to make a special claim for the rhetorical impact of confessional writing that shows an author's transition from culprit to healer, Kittredge's family story (and the stories of many people, often men) of transcending destructive beliefs and activities possesses special cultural leverage, special credibility. Another possible meaning of the title is that one not only demonstrates care for other things or people, but one receives ("takes") care—or good things—from the world and ought to show proper appreciation for this. And this appreciation for things received, the title and the text imply, is one of the functions of writing. Of course, despite the retrospective, self-critical aspect of the narrative, there is also the prospective, future-oriented assertion—the implication, for readers, that we could all learn from the author's story and take pains to be careful as we move forward with our own lives. Kittredge writes:

> Falling back into the world through a reopened gate, into a time in which we were seamlessly wedded to every thing, is a dream. I want

it to be possible; I want things to be that good. But childhood is over. Our invulnerabilities are gone. Creatures, our only companions on this planet, amid the swing of infinities, are dying out. It's time to give something back to the systems of order that have supported us, some care and tenderness. We need to give some time to the arts of cherishing before much of what we adore simply vanishes. Maybe it will be like learning a skill: how to live in paradise. (12)

To be honest, I'm not sure there's anything particularly masculine about the concepts of "looking after" and "taking care," or about the language strategies evident in the two books I've just described. However, I think it's important, even on a symbolic level, to recognize the wisdom and constructiveness of these statements by contemporary male authors—I say "symbolic" because, as is quite obvious to anyone who studies environmental literature, many of our most eloquent speakers and writers on behalf of social reform and environmental protection happen to be male. In our reflexive rush to look for culprits and scapegoats, to root out the source of social injustice and environmental ravagement, we must take pains not to disenfranchise *any* who might contribute much-needed words and insights. In his well-known essay "A Word in Favor of Rootlessness," considering the tendency of environmentalist stay-at-home thinkers to put down restless wanderers, John Daniel writes, "I feel at least a tinge of concern that we might allow our shared beliefs and practices to harden into orthodoxy, and I fret that the bath water of irresponsibility we are ready to toss out the door might contain a lively baby or two" (260). Like the orthodoxy of rootedness, the growing orthodoxy of implied or explicit critiques of masculine perspectives on nature and culture seems, at times, in danger of throwing out babies with bath water.

During the 2002 Great Basin Book Festival in Reno, I moderated a session with six western women writers and artists who were also ranchers; the issue of gender never came up. Two days later, I moderated a panel of male nature writers. When there was a break in the discussion with the men, I used it as an opportunity to ask how they felt about the standard critique of male ways of viewing nature. All three quickly pointed out that their recent and current work explicitly counteracts stereotypical male points of view.

David Strohmaier, the author of *The Seasons of Fire: Reflections on Fire in the West* (2001), said that while the traditional American attitude toward wildfire has been one of conquest and eradication, his own work argues that we are fascinated by fire and that fire has an important function in the natural world. His book shows an intimate engagement with fire, not the traditional, militaristic quest to overcome this phenomenon.

John R. Campbell quickly reminded listeners that the third and final section of his book of essays, *Absence and Light: Meditations from the Klamath Marshes* (2002), is called "Intimacy" and explores various forms of physical and psychological connection among humans and between humans and the rest of the world. In other words, his lyrical meditations are an implicit response to the famous passage from the "Burrowing Owls" chapter of Terry Tempest Williams's *Refuge*, in which Terry's friend, Sandy Lopez, states, "Many men have forgotten what they are connected to.... Subjugation of women and nature may be a loss of intimacy within themselves" (10).

For his part, poet and essayist William L. Fox, author most recently of *Playa Works: The Myth of the Empty* (2002), spoke about his recent stay in Antarctica as part of the National Science Foundation's program for artists and writers. In particular, he told the story of an excursion he took one day with a woman scientist, in search of a patch of moss on the ice. The scientist explained that men often navigate on the ice by using GPS technology alone, ignoring the physical details of the frozen landscape they're passing through, while women tend to look closely at the patterns in the actual ice and, as a result, have a history of finding their way more effectively through this isotropic landscape. "We found the moss, a tiny patch in the midst of a landscape of ice," Fox recalled. "And I learned a new way of seeing that place."

This is what it all boils down to, isn't it? Learning new ways of seeing, thinking, expressing ourselves, behaving. Perhaps it doesn't matter if a writer or a scholar is a man or a woman; perhaps it doesn't matter if a person is expressing traditionally male or female attitudes toward the world. To the extent that ecofeminism has helped to sensitize us to certain healthy viewpoints and to destructive viewpoints, it's all to the good. To the extent that a new, provisional "ecomasculinism" might help us avoid essentialist castigation of all men and all masculine attitudes, that's useful too. But in the end, I think, we would be wise to transcend both ecofeminism and ecomasculinism and concentrate on taking care, regardless of gender, to make our presence in the world as benign as possible.

Works Cited

Bass, Rick. "The Farm." In *Getting Over the Color Green: Contemporary Environmental Literature of the Southwest*, edited by Scott Slovic. Tucson: University of Arizona Press, 2001.

Campbell, John R. *Absence and Light: Meditations from the Klamath Marshes*. Reno: University of Nevada Press, 2002.

Clark, Suzanne. "Annie Dillard: The Woman in Nature and the Subject of Nonfiction." In *Literary Nonfiction: Theory, Criticism, Pedagogy*, edited by Chris Anderson. Carbondale and Edwardsville: Southern Illinois University Press, 1989.

Daniel, John. *Looking After: A Son's Memoir*. Washington, D.C.: Counterpoint, 1996.

———. *Winter Creek: One Writer's Natural History*. Minneapolis, Minn.: Milkweed Editions, 2002.

———. "A Word in Favor of Rootlessness." In *Literature and the Environment: A Reader on Nature and Culture*, edited by Lorraine Anderson, Scott Slovic, and John P. O'Grady. New York: Addison Wesley Longman, 1999.

Dillard, Annie. "To Fashion a Text." In *Inventing the Truth: The Art and Craft of Memoir*, edited by William Zinsser. Boston: Houghton Mifflin, 1987.

Duncan, David James. *My Story As Told by Water*. San Francisco: Sierra Club, 2001.

Elder, John. *Reading the Mountains of Home*. Cambridge: Harvard University Press, 1998.

Fox, William L. *Playa Works: The Myth of the Empty*. Reno: University of Nevada Press, 2002.

Gaard, Greta, and Patrick D. Murphy, eds. *Ecofeminist Literary Criticism: Theory, Interpretation, Pedagogy*. Urbana and Chicago: University of Illinois Press, 1998.

Hardin, Garrett. *The Ostrich Factor: Our Population Myopia*. New York: Oxford University Press, 1998.

Hay, Peter. *Main Currents in Western Environmental Thought*. Sydney: University of New South Wales Press, 2002.

King, Ynestra. "Healing the Wounds: Feminism, Ecology, and the Nature/Culture Dualism." In *Reweaving the World: The Emergence of Ecofeminism*, edited by Irene Diamond and Gloria Feman Orenstein, 106–21. San Francisco: Sierra Club Books, 1990.

Kittredge, William. *Taking Care: Thoughts on Storytelling and Belief*. Minneapolis: Milkweed Editions, 1999.

Kolodny, Annette. *The Land Before Her: Fantasy and Experience of the American Frontiers, 1630–1860*. Chapel Hill: University of North Carolina Press, 1984.

———. *The Lay of the Land: Metaphor as Experience and History in American Life and Letters*. Chapel Hill: University of North Carolina Press, 1975.

Lindholdt, Paul. "The Spray and the Slamming Sea." In *On Nature: Great Writers on the Great Outdoors*, edited by Lee Gutkind. New York: Tarcher/Putnam, 2002.

Lopez, Barry. "A Scary Abundance of Water: Growing Up with the San Fernando Valley." *LA Weekly* (11–17 January 2002): 26–34.

Nabhan, Gary Paul, and Stephen Trimble. *The Geography of Childhood: Why Children Need Wild Places*. Boston: Beacon, 1994.

Plumwood, Val. *Environmental Culture: The Ecological Crisis of Reason*. London and New York: Routledge, 2002.

———. *Feminism and the Mastery of Nature.* London and New York: Routledge, 1993.

Sanders, Scott Russell. *Hunting for Hope: A Father's Journeys.* Boston: Beacon, 1998.

Simonson, Harold P. *Going Where I Have to Go: Essays from Within.* Portland, Oreg.: Strawberry Hill, 1996.

Slovic, Scott. "'Be Prepared for the Worst': Love, Anticipated Loss, and Environmental Valuation." *Western American Literature* 35, no. 3 (fall 2000): 236–63.

Strohmaier, David. *Seasons of Fire: Reflections on Fire in the West.* Reno: University of Nevada Press, 2001.

Warren, Karen J. "Feminism and Ecology: Making Connections." *Environmental Ethics* 9 (1987): 3–20.

Williams, Terry Tempest. *Refuge: An Unnatural History of Family and Place.* New York: Vintage, 1992.

THE "NEW MAN" IN NATURE

Anecdote of the Car
The Diminished Thing

ALVIN HANDELMAN

I saw the wreck for the first time in 1996. It was a half mile from our cabin, down the Old Stagecoach Road, an abandoned track that once connected West Newbury to East Corinth, two back-country Vermont towns. The sky was cloudless. A light breeze lifted the maple leaves. I could hear hermit thrushes singing their sweet, melancholy songs. I walked beside a farmer's old rock wall, now tumbled boulders; his once grassy sheep meadow now pastured a jack pine forest. Far up the road was something new, bright and metallic, but at this distance unrecognizable. It was a yellow Toyota sedan, rolled on its roof. The windshield and rear window were smashed. The doorposts were bent like elbows. The seats hung upside down and were covered with a white mold. Pipes and tubes lay exposed in the undercarriage. The tires looked like the paws of some grotesque cartoon creature, playing dead. Broken glass and detached pieces of chrome sparkled in the grass all around. In a segment of shattered headlight reflector, I saw my funhouse face, surprised and chagrined, and behind it a white cloud.

Since that first shock, I've seen the car more than a dozen times in the past four years. Oddly, the shock does not diminish. I prepare myself for it the way a high jumper reviews the individual stages of the jump. "The car's just up ahead," I tell myself. "It's a yellow hulk, rusting, chrome and glass all around—." Then I'm in front of it. It still dumbfounds me. It's always worse than I've imagined. The juxtaposition of leafy forest, rutted road, singing thrushes, and flipped, smashed junker is too great to compass. It almost rises to art—Dada by Duchamp.

I can guess who hauled the car here. Not artists, but dumb-drunk yahoos. Before they trashed the area, I'd walk up the road knitting Frost poems together. I'd start with "The Road Not Taken": "I shall be telling this with a sigh." I'd go to "Mending Wall": "Something there is that doesn't love a wall." I'd jump to "Come In": "As I came to the edge of the woods, / Thrush music—Hark!" I'd conclude with "Directive": "Make yourself up a cheering song of how / Someone's road home from work this once was, / Who may be just ahead of you on foot."

Now the walk is Wallace Stevens's "Anecdote of the Jar":

I placed a jar in Tennessee,
And round it was, upon a hill.
It made the slovenly wilderness
Surround that hill.

The wilderness rose up to it,
And sprawled around, no longer wild.
The jar was round upon the ground
And tall and of a port in air.

It took dominion everywhere.
The jar was gray and bare.
It did not give of bird or bush,
Like nothing else in Tennessee.

But unlike the jar that "[makes] the slovenly wilderness . . . no longer wild," the car turns the wilderness into a dump. It's an assault. On aesthetics and the emotions, but also on simple logistics. Why would anyone go to such trouble? The Stagecoach Road is eight miles from town. The last two miles are dirt. The rise in elevation is more than eight hundred feet. For all their hard work, they might have towed the car to a junkyard and gotten a few bucks for scrap. Instead they hauled it up here. I can see them dragging it in, pulling it on a chain behind a pickup. I imagine them rocking the car to tip it over. It goes up on the driver's side, poises for an instant. One of the guys slips in the mud in his hurry to get out from under, as the car finally rolls completely over. The roof buckles; the windshield breaks. They laugh at their mud-covered buddy, his shit-eating grin. It's the funniest thing they've ever seen. The work done, the laughter over, the six-packs finished, the buzz gone, they climb back into the truck and drive home to sleep it off.

Once Were Yahoos

Though I don't own the Stagecoach Road or the land around it, I felt proprietary and took this assault personally. It was not, to my mind, an isolated incident, but the culmination of a series of events, all of them beyond my control, most perfectly comprehensible, a few not. I first moved with my friend Eric to this area thirty years ago. We became caretakers of a farmhouse called the "Old Perkins Place," one of three houses on Wright's Mountain Road. There was a house at the bottom of the road, then the Perkins house a mile up. Three-quarters of a mile farther on, a fork in the road went left to the Appleton farm and beyond, or right, uphill, passed the Stagecoach Road at the height of land, then downhill back into town.

Eric and I arrived in December. Cold and snowy. Wright's Mountain Road was a bobsled run, two berms of snow on either side of the skiddy track, barely wide enough for two cars. The Perkins farmhouse sat below a white hill, crowned by maples. In front of the house stood an enormous old elm and a mature butternut. I had come by way of Cambridge, Providence, and New York, where I had grown up and gone to college; Eric by way of the Peace Corps (Tunisia) and Long Island. I sought connection with the land. I wanted permanence, a quiet place to live and write. I thought I had found it. Wright's Mountain Road, the Perkins Place, the land around us were perfect and lovely. I figured the place would stay this way. The elm might get Dutch elm blight, the butternut might come to grief in a windstorm, but Wright's Mountain and the houses, hills, meadows, and forests would stay as they were. This seemed a reasonable assumption given the history of the area and its distance from the usual attractions that bring people and development. The landscape was simple and spectacular to me, but nothing approaching Albert Bierstadt sublime. Wright's Mountain was 1,836 feet high, a high hill, really. There were no ski lifts or cross-country ski trails. The nearest major lake, resort, or river lay eight to ten miles away. This was a land of Holsteins and heifers peacefully grazing. Interstate 91 had opened in 1963 but had not brought in thousands of tourists or vacationers. There were no major businesses within thirty miles. The population and land prices around Wright's Mountain Road had been in decline for fifty years. The clincher, I thought, was the road itself—a steep, dirt track that went (not necessarily in this order) from dirt to snow to ice to mud, and back to dirt again. In summer the dirt became a washboard that rocked your kidneys. In winter, if it snowed more than six inches at one time, you stayed home, because if you drove down, you wouldn't make it back up. The electricity went out at the least provocation. You shared your telephone line with strangers. The phones went dead if you looked at them askance, which you did often because it was so difficult dialing out. This place was made for would-be writers, but not for city folk who had to get to work on time.

Most city people wouldn't want to live here, I thought. It was too primitive, too rough, especially in winter. The Perkins house had a wood-burning stove in the basement that had to be fed chunks of oak every three hours. If we were lazy and didn't get up and trudge down through the cold house from our upstairs bedrooms to feed the fire, the house was freezing in the morning. Eric and I were used to roughing it. We had hiked and camped out in the mountains in New Hampshire and Maine for years.

Unlike Boston or New York, this place and landscape would not be in flux and fuss. This was a place of granite, which in every meadow showed its knees and elbows. This place would endure. I fell in love. Actually, back in love. As a camper and counselor, I had spent ten sweet summers at a camp in West Fairlee, ten miles away. I was connected to this area by hundreds of associations that spawned hundreds more, from my first banana split to my first kiss. There were enough memories here to make the place home before we settled in. What better spot to learn the names of the birds and flowers, to take walks, to write short stories and novels? There was no radio, no TV, no telephone (that first year). We heard only the yank-yank of a hungry nuthatch in the elm, the hopeful "spring soon" of the chickadees. The whir of grosbeak wings at the feeding station. We were young, twenty-seven, good friends sharing the same dream—to be good writers. I saw so much beauty all around me I wanted to shout. I heard what I imagined was the music of the universe. What else could a silence so vast and essential be? I could deal with the minor difficulties and privations and could live on very little money. We were house-sitting, rent-free, for an architect in Cambridge who had bought the house and eleven acres for next to nothing. Dick came up on weekends and brought our friends. We played guitars, sang songs, got stoned, made love, and, unwittingly, scared the neighbors.

Slippery Slopes

Our neighbors were as unlike us as I imagine I am to those yahoos and their junk car. Not upstarts or fools, our neighbors were old-time Vermonters, many of whom had been born and brought up in the houses they were living in. They were small-herd dairy farmers, with ten to fourteen head. They sold milk, raised crops for canning. They killed a steer, a pig, and a deer or two each year for meat. They kept chickens for food and eggs. They lived close to subsistence. They had heard about the troubles in the cities, about hippies and drugs.

Innocent of country life, we at first did not grasp that we were living in a tight-knit farm community whose people were married to the land and the seasons. Two hundred years of traditions were being passed along from fathers to sons, mothers to daughters. Our neighbors, we learned eventually, had gone to one-room schoolhouses. Many had only eighth-grade educations. They knew everyone within thirty miles. Here, you were either a farmer or you worked in a farm-related industry. Our neighbors knew tourists—summer-camp personnel and summer people—up for a few weeks, a month, then gone. Summer and winter, most city folks kept to the Connecticut River Valley and did not venture into the hills, where there were nothing but farms and dirt roads that forked off in odd directions, obedient only to topographic logic. Yet we had plunked ourselves down among them. To them, we were very odd. They did not think gay, not in 1969, and we were too young to be called "old bachelors." They knew for certain we were not Vermonters, not farmers, not country people. Our hair was long. We didn't shave every day. Apparently we didn't work. We were as dangerous to the people here, and to their environment, as that busted up junker was to me. We started diminishing the place the moment we arrived.

What did we do? We took long walks in the woods. We learned to recognize balsam fir, spruce, hemlock, elm, hickory, ash, poplar, beech, and several types of birch, oak, pine, and maple. We birded: chickadees, hairy and downy woodpeckers, crossbills, pine siskins, brown creepers, redpolls, white- and red-breasted nuthatches. Evening grosbeaks came in flocks, flashes of Joan Miró. We saw one shrike, its prey hung on a twig, just like a picture in a book. We met no one. No neighbor came by with a hot dish to welcome us to the community. We wrote. We read books on nature. It didn't occur to us how odd we were, not only to the Vermonters but to most other people. American males at our age, with our education, were already well-ensconced in good jobs, married and settling down, or on the verge of both. Yet to us what we were doing seemed a perfectly natural result of our readings in American literature, our camping experience, and the nationwide zeitgeist—evident to us, at least—to get back to the land.

A month after we arrived, we walked up the road, passing a half mile of dark forests and snow-filled fields. We walked by the barn and house of our nearest neighbor, whose plaque on top of the mailbox read "Appleton & Sons." We ignored the barking of a frenzied and indignant dog and the none-too-subtle gawkers behind the curtains at the windows. What a sight we must have been that cold, brilliant January day. We were wearing fat white Arctic boots—"Mickey Mouse boots"—to keep our feet from freezing, wool watch caps, and long wool army coats. Binoculars dangled from our necks. The Appletons, we learned later, knew we were the flatlanders who lived at the Old Perkins Place, purchased from Bob Pierson for ten thousand dollars. Pierson had bought the buildings and 410 acres from Howard Perkins for the same price eighteen months earlier. We learned that this sale and "quick" resale had left people in the area scratching their heads. Who but fools paid ten thousand dollars for eleven scrub acres, an old barn, and a decrepit house, in which the kitchen floor sloped at a seven-degree angle from the table in the corner to the sink opposite?

A quarter of a mile further on, we came to a mailbox marked "Tuckers" in bold black letters. The house nearby looked ramshackle. From a tree limb in the front yard hung a red fox. We were shocked. Appalled. This confirmed my suspicions about our neighbors, suspicions I had only recently developed (nonetheless now hastily applied) from reading Thoreau's journals and John Stewart Collis's *The Triumph of the Tree*. These were manure-slinging farmers, New England Puritans, who loved their God but hated most of His creations. They were narrow-minded and self-protecting. They refused to see the larger picture, how nature was a chain of creatures, each with its important place in the Great Cycle. If we consistently killed predators, we would soon be overrun by their prey. I believed we must live in harmony with God's creatures. And what better place to do that but in this bucolic wonderland? Where humanity was but a warm spot in a sea of snow. Where only two vehicles, going and coming back, passed on the road each day: a red VW and a big yellow school bus. Weekends were eerily silent. After sunset on a cold February day, I stood in the middle of the road and gazed at its slow rise between the tall bordering maples and its graceful sweeping turn to the left. Above the dark limbs of the trees, the clouds were gunmetal gray, and I felt a deep, overwhelming sense that the road went on into a vast and inviolable wilderness, and that I was part of it, part of the sky, the trees, the road, the perfect, all-embracing solitude, part of a natural, immutable wholeness.

Despite the idyll, I wasn't ready to settle down. In May I moved to Boston to earn some money, then moved into a commune in Albion, near Mendicino, California. The redwoods were magnificent. The blue Pacific and its rocky shores were spectacular, but nothing there felt right. Maybe it was the warm November rains, or the high angle of the sun, or the three lesbians I was living with. Whatever the reason, I felt out of place, out of time, out of season, and entirely superfluous to those women.

Dominions

Seven months after leaving Vermont I came back. I wasn't gone long, yet the place I had thought was immutable had changed. On top of the gray cliffs of Wright's Mountain someone had built a cabin. We could see it plainly from the porch of the Perkins farmhouse. It was small and red and had a gray tin roof. It squatted up there. You couldn't miss it. Its presence spoke "house" and "people" where there had only been trees and rocks. It changed the feel of the place—like the jar in Tennessee, the car on the Stagecoach Road, and the two of us in the house. It took dominion. At times I imagined that by leaving the farm I had allowed this to happen. I had prayed for permanence, then run off to, of all places, California. I got what I deserved. Some yahoo had built a cabin on the mountain.

On my first walk, I climbed to the top of the hill behind the house through the blue-white snow. To the east the cabin seemed to jut out over the landscape. A trompe l'oeil of perspective, a spatial trick that was only a bit more annoying than the odd whining sounds in the distance. Mosquitoes in mid-winter? I wondered. I walked to a spot where I could look north toward the Appletons' farm. Tiny forms were moving across the distant meadows. Skimobiles! Jeez! What fools, I thought. Didn't they realize they were destroying the peace and quiet? Didn't they know what their machines did to the bushes and saplings? Snowshoes and cross-country skis were good enough for me and my friends. Why weren't they good enough for those yahoos?

A week later, I met Ernie Appleton, the man responsible for the skimobiles, who was, I imagined, in the arrogance of youthful self-righteousness, ruining this place. I was walking behind the house, seeking, as I wrote in my journal, unaware of my preciousness, "the wonder of the winter landscape, the silent solitude." He roared up to me. I had met him briefly on a dark, stormy night the year before. The odor of hay and manure and barn animals had wafted from his red and black wool coat as I opened the door. I couldn't make out his features in the darkness. He was tall and spoke with a thick Vermont accent as he told me his name and said he was my neighbor, up the road. He said his phone was out. He asked if ours was working. The phone, visible in the living room, wasn't hooked up. Eric and I kept it around for a joke. I told him we didn't have a phone.

Now in the brilliant afternoon sunlight, I could see Mr. Appleton clearly. He was scarier than he had been that night. He was wearing a red stocking cap with black stripes and the same wool coat. Luckily, the xenophobia of *Deliverance* and its movie was still a few years off. At best, he was an escapee from a Dr. Seuss book. At worst, a hard case on the lam from Sing Sing or Bellevue. He was kneeling on his right leg. His left leg trailed along the running board of his loudly combusting machine. A jaunty riding stance, I thought. He grinned at me. Not a tooth in his mouth. He reminded me of the men I avoided in New York—panhandlers, derelicts, schizos, all with socko breath, all in various states of disintegration. They scared me because they were men who did not play the usual roles, or had fallen out of them. I did not feel terribly secure in mine. What's a writer if he isn't writing, or stops writing? He's a bum. I could end up there too. I had to be scowling. In front of me was a dislocated Bowery bum, riding a clattering skimobile.

"Met you last winter," he said. "Ernest Appleton."
I told him my name.
"Where you from?"
"New York City," I said, not without pride.
"You like it here?" he asked.
"I love it." I said. Then I shouted, "Especially the peace and quiet!"
"How do you like my cabin?"
"That's yours?" I said.

ANECDOTE OF THE CAR

"A-yup." he said.

"Huh," I managed to say. So, he was responsible for that too!

"Wal," he said. "Pleased to meet ya." He shot off, corrugating the snow behind him in a narrow sweeping curve. I discovered that this track made walking in the deep snow significantly easier. Despite this somewhat perplexing benefit and the oceanic silence that descended minutes after he disappeared, I cursed the man and his machine and his cabin. To me these were definite blows to the environment, to the ambiance of the place—real diminishments. I could not imagine, then, that these developments might serve as metaphors for the changes we undergo, within us and around us, as we grow older. That wasn't within my ken at twenty-eight. Yet the changes were there, done, beyond my control. I could storm and shout. I could move away.

I stormed and shouted. Ten days later, Appleton stopped to ask if he could use the abandoned sugar house on the property to make maple syrup. We said sure. He drilled the trees along the road and on the hill behind the house and hung sap buckets from the spiles. Over the next two months, Eric and I, and Mr. Appleton, as we called him at first (his bearing seemed to demand it: he was six feet six inches and very self-assured), collected the sap, boiled it down, and made maple syrup and candies. By the end of that distilling process, we were becoming friends. To us city boys, nature dreamers, a new world had opened up—country life. Ernie Appleton had a lot to teach us about plowing a straight furrow, raising a garden, hunting deer, making hamburger, getting along with your neighbors. He was unlike anyone we had ever met. He had no qualms, for instance, asking you to walk behind the manure spreader to check the evacuation mechanism, while it was running. If in your naiveté you got covered with cow shit, he'd laugh at you, tell you to go home and wash up, and hurry back to help him with the next load. At the 1971 annual meeting of Central Vermont Community Action, which was attended by state council members and representatives, Ernie put on a burnoose and fez that Eric had brought back from Tunisia. He had Eric teach him the worst Arabic curse he knew—*ziporoomic* (your mother's cunt). At the meeting, Ernie greeted everyone with a salaam, a rolling hand gesture, and a hearty, heart-felt *ziporoomic*. "*Ziporoomic*, Representative Moran." "*Ziporoomic*, Congressman Peters." "*Ziporoomic*, Dr. Stone." Smiling broadly, each of his acquaintances nodded and returned his salaam. Ernie liked to stir things up. He was restless, curious, and fearless. He was a mackerel in a holding tank. He was also forgiving. He had seen our phone the night he came down to ask to use it. He thought we were cheap sons of bitches. He had heard the rumors about us too. Our neighbors told him we were long-haired hippies. Pot smokers. We had no furniture in the house. We slept on bales of hay. Girls came to the house. Different girls every weekend. He ignored the gossip and made our acquaintance. He was the embodiment of old New England, that place of subsistence, hardscrabble farms, of getting by, making do, doing without. He was not pent and closed-minded and bound by his traditions but bolstered and emboldened by them. He knew Frost and Jewett by heart. He had mended wall and cleaned out the pasture spring. He put on his long johns the first day of September and took them off the last day of April. He cut, hauled, and split six to seven cords of wood each fall. He judged men by how they handled an ax. We knew no one like him, no one—certainly not our fathers—who had such a profound understanding of the landscape and could make it yield firewood, meat and vegetables. Luckily for us, he wanted to know who we were and what we were doing here. He wanted to see if the rumors were true. He had seen us in our winter garb, binoculars dangling, and figured we couldn't be very dangerous. His wife, Sylvia, more cautious, admitted that when we walked past their house that first time, she thought we had escaped from the insane asylum in Waterbury. She called their ten-year-old daughter, Laurie, into the house. She then turned to Ernie and, innocent of Swift, said, "Look at those two yahoos!"

Ernie put country life in perspective. Puffing Prince Albert tobacco on his corncob pipe, he said the red fox we had seen at Tucker's had killed nearly all Tucker's chickens. He himself shot deer and sicced the dog on woodchucks because they ate the vegetables in his garden. He needed those vegetables during the winter. He couldn't afford to buy vegetables at the supermarket. He hunted deer in season (and sometimes out) because he liked to hunt and needed the meat. "Do you boys think deer are cute?" he asked. "Have you ever hunted? Have you ever needed to hunt? Venison's awful tasty," he said with a big, sly smile. He said his father talked about seeing a wolf in 1918. But maybe it was a big dog. Mountain lion? They were gone long before he was born. He said bear were nearly gone now. He had walked and lumbered his four hundred acres for fifty years. He had a keen eye, but he had seen only two bears in all that time. There were no coyote, and he had never seen a bobcat, though he had heard them howl. No moose either. They were in Maine. He raised pigs and cows. "You want to learn how to make hamburger?" he asked.

Ground Beef

It was a slate gray day in late fall, the air sharp and damp, dry dead leaves awhirl. Ernie and Gallerini, the butcher, went down to the meadow behind the barn to fetch Betty. She sensed something was wrong and bolted. Ernie walked after her, unhurried. Gallerini hung back, the rifle at his side. I watched from the ramp leading into the barn. Ernie called his old cow. She had given him years of milk and calves. He didn't like this business. He always made the mistake of caring too much for his cows. He tried to hide this from the cow and Gallerini. It wasn't the manly thing. Betty nibbled her last supper. Ernie picked up her halter and led her to the barn. The cow mooed placidly, nuzzled him in the small of the back. She wanted to know who the stranger was, and what was up. At the entrance to the mow, Gallerini aimed his .22 at Betty's thick skull and fired. She mooed loudly, twisted her head, wiggled her ears, stuck her tongue in her nose. Leaning into it, the butcher aimed again, the muzzle inches from the white curly locks on Betty's forehead. Betty was going down as the explosion rippled outward, echoing off the mountain. She did not twitch. She did not position her legs as her big body did death's pratfall. Gallerini gently lifted her chin, as if he were a barber about to give her a shave. He slid his knife along her throat, unzipping the skin, and the blood poured out over the granite sill in silken crimson waves.

Ernie and Gallerini pushed Betty over on her side, careful not to get caught beneath her. They shoved steel hooks through the flesh between the tendons and bones of her hind legs. They attached the hooks to a wooden spreader and fastened the spreader to a cable. The cable ran up through a pulley hanging from a beam above the barn doors. Ernie connected the cable to his tractor. He drove forward slowly. The cow began to slide backward on its side, its hide scraping the barn's floorboards. Gallerini worked quickly with a big knife, cutting with deft, vigorous strokes. A moment later he placed the cow's head next to the door. Its eyes were clear; its ears flopped fetchingly, its mouth slightly ajar. Its expression showed neither sorrow nor surprise, neither regret nor fear. A perfect, indecipherable insentience.

The cow's hind legs formed a Y on the spreader as Ernie drove forward. The carcass rose into the air. The beam took the weight, creaked and groaned. Moments later Betty was performing a clown's headless handstand. Then it was free of the ground entirely. Ernie set the tractor's brakes and clomped back to Gallerini's side. They worked at the hide, yanking it down, ripping it off the subcutaneous flesh as if it were an enormous black and white glove. Together they stripped the hide to the forelegs and cut it off. They lugged the thick skin out of the way, dropping it near the head. The eyes were glazed over now, the luster gone. But the huge dangling body, pink and white, still seemed alive. Tiny muscles twitched everywhere, synaptic transmissions in the right hind leg, in the left foreleg, in the neck, in the udder. A last minute alarm—danger! flee!—sent too late. It reminded me of the night I stopped on a vast flat plain in Nevada, on the way to California, and watched heat lightning go off in different quadrants of the sky.

With a crescent-shaped knife, Gallerini slit the belly open in one long motion, top to bottom, a labial incision that exposed the guts. A few quick slices and the great tubal mass of pink intestines disgorged onto the floor. A crazy birth. They dragged the slimy cords of offal across the hay-strewn floor and dumped them near the hide and head. A few flies with big, iridescent bodies were already at the carcass. Ernie and Gallerini now began the more delicate work of separating the urinary and reproductive tracts from the meat. Like Adam, Gallerini named the parts, making unsubtle references to sex and a woman's body. Finally, the butcher began the transfiguration from flesh to food. He created the steaks, loins, ribs, chucks, shanks, briskets, flanks, rumps, and the rounds that when grounded became hamburger. A week later, Ernie stopped at the house and handed us a pound and a half of hamburger, neatly wrapped in white paper.

DIMINISHED THINGS

There is a singer everyone has heard,
Loud, a mid-summer and a mid-wood bird,
Who makes the solid tree trunks sound again.
He says that leaves are old and that for flowers
Mid-summer is to spring as one to ten.
He says the early petal-fall is past,
When pear and cherry bloom went down in showers
On sunny days a moment overcast;
And comes that other fall we name the fall.
He says the highway dust is over all.
The bird would cease and be as other birds
But that he knows in singing not to sing.
The question that he frames in all but words
Is what to make of a diminished thing.

Ernie Appleton held dominion over Wright's Mountain Road for fifteen years. The man who had built the cabin and introduced skimobiles had also introduced us to a world rich with characters, legends, and humor, with stories of people we city boys had never known, with a way of life that was slowly dying out. In 1985, Ernie died of lung cancer. He had encompassed the neighborhood, not only because he owned a large portion of it but because he knew every living and historic aspect of it. Then came the land run. To pay real estate taxes, Sylvia was forced to sell five parcels, one included a three-hundred-acre tract of Wright's Mountain, which was purchased, fortunately, by the town's nature conservancy. Pierson built a house on land he had bought from Perkins and sold three ten-acre lots. Others were busy selling, too. Still others, building.

Every summer, my wife, Carol, our son, Matt, and I returned from what I considered exile in the Midwest to find that a new house or trailer had risen where once there had been a field or forest. Suddenly the hundreds of acres I had cherished for walks—to be alone, far away, to transcend time, or, more simply, to watch birds, identify flowers, sight animals—those lands were now closed off by no-trespassing signs, fences, and houses. Cars? Where once there had been two vehicles going down and up, now there were dozens, and all in a hurry. The changes I observed as I returned every July were not incremental, but monumental. It was like running into a teenager you haven't seen for a year who's shot up ten inches, gained twenty pounds, and dyed his hair purple. And the next year, he's taller and heavier than you, and blotched with acne. By the summer of 1996, there were twenty-two buildings on the road where once there had been three.

I sought an explanation. Had the population of Vermont increased exponentially? No, it had been half a million for years. Had construction exploded on the two roads parallel to Wright's Mountain Road? I drove up and down both roads, counting new houses. There were only minor increases. Our road was an unfortunate anomaly, and a number of factors accounted for it. Pierson's business acumen. Sylvia's monetary need. Word of mouth. Bad luck. Maybe God played a hand, too. At times I imagined this was God's unique and brutal response—definitely an Old Testament response—to my prayer to leave this place intact. I wanted stasis and solitude. Instead, the road became chockablock with houses. Then I took that walk down the Stagecoach Road and saw the clunker—it was the coup de grâce.

Bear, Jar, Car, Bird

I thought the landowner would have it removed. He didn't. I thought I'd get over the shock. I didn't. Then I thought I'd sell the cabin. Move to a less populated area in Vermont's Northeast Kingdom or Maine's northern tier. Head for the frontier. Go West, Young Man (or old). Alaska! These would have been manful responses to the problem. Lighting out for the territories with my family, whether north or west, I'd have been in good company, good literary company, too. It's the American tradition, a man's way of dealing with the problems created by apparent overcrowding, unpleasant change, and discomforting diminishment. If you don't like something, you can change it. If you can't change it, you can always pack up and leave. I was tempted, but the place and the roots I had sent down tripped me up.

There are our neighbors, a motley blend. We have Jehovah's Witnesses, hard-line Catholics, born-again Christians, Baptists, and erstwhile hippies who now play golf. (In the summer we have a Jew and an Episcopal minister.) Given the intensity of their beliefs, the year-round residents might have started several small holy wars. Instead they've banded together in an alliance of mutual understanding and cooperation. Every summer they quickly fold us into their lives. They have kids, so Matt has friends. And there's been plenty to teach Matt about how to handle an axe and a chain saw; how to fell trees, buck them up, split wood, and build fires; how to handle a gun and live without running water, indoor plumbing, and electricity—amenities we don't have.

There's also a wonderful, seemingly contradictory resurgence of nature: sightings of coyote, moose, and bear have become common. Reports of wolves and mountain lion are rare but have begun to occur. In the 1850s, Thoreau elegantly mourned the loss around Concord of these animals, likening them to torn-out pages of a great book and assuming they were gone forever. Yet 150 years later, these pages, here at least, are binding themselves back into place. This revival appears contrary to all logic. People and their speeding cars usually have a deleterious effect on the flora and fauna. But the people on the road now are not Vermont farmers. They are not busy clearing forests to make pastures for their livestock or cutting a half dozen cords of wood to heat their homes. The forests have returned. (Today, 80 percent of Vermont is forest. In 1930, only 20 percent of the state was forest.) All those trees make good cover and browse. Animal ranges have expanded accordingly. Laws protecting certain species have also had a beneficial effect, as has the gentrification of the population, which tends to be less biblical in its approach to lording it over earth's creatures, and less driven by real necessity.

In more than thirty years of walks, I had never seen a bear track or scat, though I was always looking for signs. In 1999, I was well past that junked car—a shock, as always—a mile up the Stagecoach Road, when I came across a print in the mud. My hand fit comfortably inside it. It was as big as a boxing glove. Bear! What did the surrounding wilderness do? It rose up to it. I'll tell you. Enormous and wild, that print took dominion.

That same year, we watched a leggy yearling moose run out of the woods and, confused, trot beside our car as we drove by. A week later, in the same area, we saw a cow moose and her bumbling calf, knee-deep in the swamp, chewing placidly. The following year we went hunting for moose. (In 1969, hunting for moose would have been as dumb as hauling a beater up a road and rolling it over on its roof.) Carol, Matt, the dog, and I were well beyond the spot where I had seen the bear track. As we neared an area where we had seen moose scats, we began to whisper as if we had

ANECDOTE OF THE CAR

entered a library. We discussed strategy. I thought we should keep the dog behind us. If he ran ahead, rustled up a moose, and scared it, it might charge. The dog would race back to us for protection. The moose would follow and we'd be "in trouble," I said, though I was thinking "mincemeat." Meanwhile the dog trotted twenty yards ahead, and Matt said, "There he goes." Moose-conjuring words. There it was—dark and big, and twice the size of the discombobulated offspring standing beside it, a baby only a mother could love. The dog quietly trotted back to us. And now we six animals stared at each other, and time spread out before us, and all around us. For several long seconds we were in Eden, and we animals had no reason to fear one another. Then she began to high-step away. Like a brown shaggy giant tiptoeing through a field of tulips, she led her calf through a sea of ferns, climbed the hill, and disappeared.

All this resurrected fauna turned Stevens's poem on its head: I placed a bear, a moose, a coyote in Vermont, / And wild they were upon a hill. . . . These animals take dominion everywhere. They make the wilderness wild again. Slovenly? Not them. Not for an instant. What was slovenly was the fear that crept along my spine as I stared at a moose and realized I hadn't climbed a tree in fifty years.

On the way back from the encounter with the moose, we ran into the car. Matt, Carol, and the dog walked by as if it weren't there. I grumbled but didn't stop walking. I can stop now to consider what that heap might mean, four years and too many run-ins later. If nothing else, I have it to thank for setting me off, forcing me to sum up thirty years of living here— the changes and apparent diminishment that has occurred and what to make of it, finally. There's no doubting the place has changed. For the worse, I'd have said when I was twenty-eight years old. Even at sixty, I'm not happy with many of the changes, and I fear more will come. For instance, a subdivision of two dozen houses across the road from our cabin. That could happen, but no sense pulling up stakes just yet. If there's one thing I've learned living here, it's the mutability of nature and individuals. "If you don't like the weather, wait five minutes," the old-timers like to say. Wait thirty years and there's a moose grazing in your backyard, and you're watching it with your wife and kid, who yesterday was five years old and now is eighteen. I couldn't have imagined any of this when I was young and hoped I could hold my world in a perfectly poised and timeless balance, a world that was simply, incredibly beautiful. Luckily, happily, that bit has not changed. Despite all the transformations, inward and outward, the landscape is still as serene and lovely as it was. Like the people on Wright's Mountain Road, this landscape is a motley blend of meadows and forests, rolling hills, low and high. It's a landscape of human shapes and sizes—breasts, bellies, and buttocks, biceps and shoulders, elbows and knees. Pastoral stuff. A landscape on a human scale. Scanning the slope of a hill or the cut of a meadow framed by trees gives me—three decades after I first arrived here—a visceral thrill of wonder and a bone-deep sense of contentment.

A quarter of a mile down the road from our cabin there are four houses, four families, a community. On a clear day, when the breeze rises from the valley, I can hear the sounds of children playing, their reedy voices lifted on the summer air, their dogs barking counterpoint, their parents calling them home for dinner. This is the best you can make of a diminished thing.

Notes

"Anecdote of the Jar": From *The Collected Poems of Wallace Stevens*, by Wallace Stevens, copyright 1954 by Wallace Stevens and renewed 1982 by Holly Stevens. Used by permission of Alfred A. Knopf, a division of Random House, Inc.

"The Oven Bird": From *The Poetry of Robert Frost*, edited by Edward Connery Lathem. Copyright 1916, 1969 by Henry Holt and Company. Copyright 1944 by Robert Frost. Reprinted by permission of Henry Holt & Co., LLC.

Traversing the Timelines

DAVID COPLAND MORRIS

Interstate into the Wilderness

I am off to Talapus Lake for a day hike in the Alpine Lakes Wilderness Area, a trip I've taken many times. There's nothing extraordinary about this excursion—put on the Gore-Tex boots, buckle the nylon fanny pack, fill the plastic water bottles, and grab a couple of Power Bars. Then down the elevator to the underground garage, into the car, out through the heart of downtown Seattle, and on to Interstate 90. Twenty-five minutes east to the edge of megalopolis, twenty-five more up to the exit for Forest Service Road 9030, then three miles fishtailing up the gravel to the trailhead. Forty-five minutes of hiking to the small weathered plaque marking the wilderness boundary. A hundred yards farther into the wilderness lies the rocky bowl of the pristine lake, mostly unaltered by the human will.

The big bang, the formation of atoms, the coalescing of galaxies, the birth of the solar system. The cooling of the earth, the emergence of life, the development of multicelled creatures, the advent of oxygen, the evolution of mammals, the transformation of animal corpses into flammable oil. The growth of the hominid brain, the birth of language, the rise of the Cascades, the coming of the Ice Age, the melting of the glaciers. The discovery of science, the advent of literary and philosophical romanticism, the invention of steel, concrete, plastic, and the internal combustion engine. The emigration to America of my ancestors, the mutual attraction of my parents. Me.

The interstate highway system. The Wilderness Act.

I'm looking out from the glass-walled cave that is my home twelve stories up in a Seattle high-rise. From here I see to the west Puget Sound and the entire Olympic mountain range. There is always a show—often simply a muted study in grays, but sometimes, on a gray day in autumn, a somber apocalyptic tunnel opens in a bruise-colored cloud layer above the mountains. The setting sun shines through, laying a bronze blade on the charcoal water. At sundown on perfect days in July, the flat black etching of the Olympics is backlit by a luminous sky shading from bloody orange to tropical turquoise to soft pearly silver. And, on a few of these perfect days, I can stand at my window at dawn and see the sky begin to brighten behind the Cascade range to the east. The lights of the city still flicker, and in the west over Elliott Bay a huge full moon showers its pale light on the water far below me. The tips of the tallest Olympic peaks are suddenly tinged by the rising sun with rays that as yet touch nothing in the darkened city or on the moonlit water. Today, the morning sky is cloudless and shimmers a pure, China blue. For some scientific reason that I have forgotten, a combination of factors makes the blueness of the Seattle atmosphere exceptionally deep—tempera, not watercolor.

I've lived in the Northwest for twenty-five years now, but still the high romanticism of this sound and mountain vista stuns my native midwestern eyes. Suburban Chicago supplied no such transcendent visions of landscape. I have a Seattle friend who grew up in Salt Lake City taking a backyard of mountains for granted. As a teenager, his standard weekend practice was to head up into the rugged canyons at the city's edge. As an adult, he lived in Chicago for many years. He astonished me with an account of his walks through the local forest preserves, the major public green spaces in Cook County, an area of five million souls. Those walks were the most fulfilling experiences in nature that he had ever had. The very modesty of the landscape, he said, quieted his mind and sharpened his sight. For me, as a boy, those same preserves were always deeply frustrating. Escaping the completely humanized suburb, I entered the woods in search of what might be called a wilderness experience. But after a short walk, I emerged on the other side of the preserve. There, another suburban street in all its quotidian tameness gazed mildly at me, and some part of myself slackened and sagged.

Why? When seventeen-year-old Edward Abbey first saw the southwestern landscape through the open door of a freight car, he fell in love. He had never seen the desert before, but he felt it was home: "In my case it was love at first sight. This desert, all deserts, any desert" (*The Journey Home* 12).

I was thirteen when I saw mountains for the first time, in Yosemite. I couldn't take my eyes off of them. The dizzying verticality, the blinding springtime snow, the naked rock fascinated me in a way, I believe, that someone from outside the Midwest cannot really understand. My mother, sister, and I stayed in a hotel on the valley floor, but I desperately wanted to explore one of the trails described in the park guides, a trail that went up, anywhere up. I picked Vernal Falls. When we got to the trailhead it was closed because of snow at higher elevations. I was beside myself with frustration, but since there was no snow on the part of the trail we could see, I persuaded my mother to let me go and explore it, and I somehow also cajoled my eighteen-year-old sister into coming along. I climbed as if drawn upward by a magnet. I felt deliriously fine.

We reached a place where a steep snowfield covered the trail. Beyond the snow the trail continued, hacked out of the sheer canyon wall, a metal railing installed along the edge. My sister did not want to cross, but I had developed mysterious powers of persuasion. How could she even think of stopping here? We set out. Somehow we managed to get to the other side and proceed along the side of the cliff until again we were blocked by snow. The impossibly deep canyon filled with the spray of the waterfall mesmerized me. Years later, I recognized my emotions in Wordsworth's astonishing lines from *The Prelude*, evoking one of his most intense alpine experiences. His words hit the highest note of the Romantic sublime:

> ... The immeasurable height
> Of woods decaying, never to be decayed,
> The stationary blasts of waterfalls,
> And in the narrow rent at every turn
> Winds thwarting winds, bewildered and forlorn,
> The torrents shooting from the clear blue sky,
> The rocks that muttered close upon our ears,
> Black drizzling crags that spake by the way-side
> As if a voice were in them, the sick sight
> And giddy prospect of the raving stream,
> The unfettered clouds, and region of the Heavens,
> Tumult and peace, the darkness and the light—
> Were all like workings of one mind, the features
> Of the same face, blossoms upon one tree,
> Characters of the great Apocalypse,
> The types and symbols of Eternity,
> Of first and last, and midst, and without end.
>   (131)

Although my sister actually is named Dorothy, I am not sure she shared these high Romantic sentiments; she was clearly impressed but also very worried about how we were going to get back. Justifiably so, as it turned out. I know now, of course, that the snowfield we crossed on the way up was one an experienced hiker would never have attempted without an ice ax. To venture out on it was a classic novice mistake of the potentially fatal variety. And indeed as I splayed myself against the snow, I realized that I was very close to sliding to my death. I had the distinct sense that if I made one more move, I was gone. Fear seized me, but I was immersed so deeply in the afterglow of the Wordsworthian sublime that

I accepted with uncanny calm the possibility my short life might end then and there.

I did eventually manage to spider along the snow to safety. Then, feeling sudden dread and guilt, I slowly turned to see how my sister was managing. She sat in the middle of the snowfield, paralyzed with fear, crying. She felt no sublime calm. I could not reassure her that it was safe and that she would be okay. She was far from safe, and I did not know what would happen. I could not believe that I had gotten her into this terrible danger. Somehow, she made it across the snow. Miraculously, she wasn't angry with me, no doubt wearily grateful that she was still alive.

That day I discovered how intensely I hungered for the kind of landscape that radiates the Romantic sublime. I also discovered that this level of hunger is not universal. D. H. Lawrence, who is not entirely to be trusted in these matters, thinks Shelley the most "male" of writers; says Lawrence, "In the degree of pure maleness below Shelley are Plato and Raphael and Wordsworth" (71). Lawrence's list goes down, or up, several more levels, assigning to various artists their places in his gender inferno or paradiso. I suppose Wordsworth's excessive "maleness" puts him in the second-lowest circle of hell because, Lawrence opines, a man "well-balanced between male and female . . . is, as a rule . . . easy to satisfy, content to exist." Wordsworth ached for the sublime. But is the poet who wandered among the fluttering and dancing daffodils excessively male? Try selling that to a classroom full of eighth-grade boys.

The evil twin of my experience at Vernal Falls befell me in Florida, on Route 1, south of Miami, while I was headed toward the Keys. Call it the countersublime. I was then in my twenties, driving down an endless strip of motels, gas stations, junk food joints, wrecking yards, body shops. Everywhere was the shameless egotism of signs and billboards elbowing each other for attention. A squall of panic rushed toward me. After traveling for an hour or so from stoplight to stoplight without the slightest change in the manifold of sensation, I suddenly felt a kind of cosmic claustrophobia. It seemed as if I had been on Route 1 forever—I lost any sense of having known a different world. The whole earth was carpeted with ugliness like the floor of a cheap motel. If I turned off the highway, the ugliness would continue in every direction and for all time. Not real destruction or danger or poverty, just drab, cacophonous, migrainous chaos—first, last, midst, and without end.

My throat tightened—I felt as if I were tied down and being strangled by weak, clammy hands. I whipped the car off the highway—into the shock of an ordinary residential neighborhood with trees and bushes and lawns. And no signs. It was just as when you wake from a bad dream, startled, unsure of where you are. And then the relief floods in.

I am not completely a disdainful Romantic aesthete. I eat my share of junk food; I've been grateful many times to find that cheap motel; I admire the grit of small-business owners; but in that long-ago, anti-Wordsworthian spot of time, the countersublime gripped me and left a pale, indelible scar on my mind. I know there are those who embrace a postmodern highway-strip aesthetic, but I'll never share it.

"Who are we, Where are we," cried the awestruck Thoreau nearing the wilderness summit of Mount Katahdin. I am zinging along the eight-lane I-90 bridge over Lake Washington, en route to Talapus Lake to demand some answers. The surreal mounded whitenesses of Mount Ranier on my right, Mount Baker on my left, both ninety miles away—I imagine them grave, silent sentinels allowing me to pass without notice, then exchanging bemused, ironical glances.

My car hums smoothly at seventy miles per hour, three thousand explosions per minute taking place under the hood—a few feet in front of my tender, nonmetallic body. This machine is a central engine of my life, setting its tempo and its spatial limits, but about this machine I know almost nothing. I might as well believe in magic for all that I understand of how the car came to be. Oh, I get the general principles of how it works—the highly refined remains of million-year-old marine microorganisms are injected thousands of times per minute in superfine spray into precisely machined metal chambers where they are ignited, the force of which combustion thrusts down rods connected to a shaft that turns an axle that turns a wheel that turns a tire that moves the car at seventy miles per hour.

So, at a very elementary level, I get the idea. What I cannot grasp is the phenomenology of the transformation of the human mind from the hunter-gatherer state to that of even a late nineteenth-century scientist. I cannot fathom the great chain of disciplined, linear, cause-and-effect thinking that has produced modern science and technology. I am much closer to the cargo cultist who imagines the airplane a magical visitation than I am to a Boeing engineer. And I sincerely question whether that engineer's understanding goes back much further than a high school science textbook, the content of which already was the product of long, miraculous efforts of cruelly concentrated thought.

In one of his poems, James Dickey beautifully describes an orbiting NASA spacecraft as "floating on mere procedure." But that's what our whole modern civilization is doing as well—floating on mere procedure, and there's no one anywhere who understands it in all its breadth and history.

Yet, it works. Amazingly well—in the short term. In their classic book

Living the Good Life, Helen and Scott Nearing describe their earnest lifelong efforts to live simply and self-sufficiently in rural Vermont, systematically minimizing their consumption of modern industrial products. They are rightly proud of their impressive progress toward this goal. But disingenuously tucked away in a footnote is their admission that owning a series of half-ton pickup trucks had been an absolute necessity in running their small farm. They even imply an affection for these vehicles, but they say nothing about the way the trucks linked them to the dark satanic mills of Detroit and the dim passageways of Wall Street. A startling lacuna. The picture of a wheelbarrow adorns the cover of the Nearings' book; honesty requires that it be replaced by an image of Ford's colossal River Rouge plant.

I worked in a factory one summer—for a week. I quit, then drove a cab for half the money. The factory made glass jars. I stood at the end of a long assembly line by myself amid excruciating noise and stunning heat though I worked at the quietest and coolest part of the line. I stacked boxes of jars as they came down. I couldn't keep up. I was in my own Charlie Chaplin movie. Eight hours stretched to eternity. Despite the good pay, no one lasted at that job more than six months.

Interstate 90 roars toward the Cascades. Megalopolis thins to a few townhouses outside Issaquah. I have for years read the Mountaineers series of hiking guide books, bibles for those tens of thousands exploring the back country of Washington. A fierce rhetorical battle rages in the text. Everyone is exhorted to experience the solitude of the wilderness; everyone is exhorted to avoid the crowds. Blithely or willingly blind to the contradiction, the text throws its arm conspiratorially around my shoulder, directs me to its favorite secret area, and says quite outrageously, "This is just between us."

These days I need a permit just to park at the Talapus trailhead. And recently a fellow brought a boom box to the lake itself and generously serenaded us all—radio, complete with screaming ads. I'm sure he meant no harm. The other hikers and I squawked at him like crows mobbing a raptor, and he eventually turned it off. Yet, solitude remains amazingly possible for the selfish romantic. I've had Talapus all to myself some lucky days, late on weekday afternoons. Today, after all the others are gone, I'm alone once more.

Talapus is an average Cascade lake basin—which is to say an exquisite, rough-hewn composition of color and shadow, motion and stillness, forest and rock. Wilderness evaporates irony. Time, which has trailed me from the city like a furious wind, buffets my mind as it rushes by. When it passes, I am becalmed in the healing present. I sit on the shore and let the liquor of silence warm me. But darkness comes on. I am chilly, and then cold. I rise, flick on my headlamp, then follow its tiny beam down the trail.

The Ghost River of Truth and Time

The intersection rammed two six-lane highways into each other. Drivers of shiny new SUVs sat behind the tinted glass atop their rugged heavy-duty suspensions and waited through four red lights before growling into a left turn—so that they could creep about two hundred yards to the next intersection where they would repeat the process. On each corner a giant shopping center stretched out, the buildings like portable furniture rented from a caterer who would come around shortly to pack it up. Chicagoland sluburbia—and I was back home on a visit. At the moment I was on the way to a new Barnes and Noble superstore to buy a book of poems—an incongruous mission amid that sprawling prairie of commerce, but on the phone the guy assured me that they had the book. I passed through the entrance as if through an airlock. The large sealed-up windows transformed the traffic outside into a wide-screen silent movie. The climate-controlled store was a secret cave covered with inscriptions like the Lascaux caves in France. No doubt somewhere on the shelves in the deeper recesses of the store were the ciphers of Edward Abbey. Like his Cro-Magnon ancestors, Abbey tried through his art to reflect and refract the mystery of the outside world. But for him one of the great mysteries was modernity, a taming of the wild that, once tamed, became in turn a new howling wilderness.

As I left the store and pulled up to the clotted intersection, I experienced an eerie moment of double exposure as I realized that some forty-plus years before, there were on this corner three farms and a country restaurant. My family and I used to come here to dinner when I was a kid. To me, it was an escape from the oppressive, bland diorama of the suburbs. The restaurant and farms were purely ghosts now. And yet I was not even close to the current edge of the megalopolis; now, in the year 2000, it extended for miles and miles beyond.

Recently, while teaching Edward Abbey, I recalled this mild little case of temporal vertigo. From his book *Desert Solitaire*, the class read "Down the River," Abbey's lyrical, sardonic, sorrowful elegy for Glen Canyon, then moved to his ambiguous and ambivalent essay "Lake Powell by Houseboat" written some twenty-five years later. I asked the class for their reaction to the two pieces, and, collectively, they said something like: "What happened to Abbey in that late essay? We kept expecting him to come to life, but he never did."

Writing of his experience on Lake Powell, Abbey claims to be comforted to find that a juniper he had admired many years earlier—before the Glen Canyon Dam was built—still survives. And he is consoled to hear that "although the surface of the lake seems motionless, stagnant as a swamp pond, the current of the ancient Colorado River continues to flow hundreds of feet below, a ghost river still bound, as it has been for eons and epochs, toward the Sea of Cortez" (*One Life* 89). At the end of his essay, taking account of all he has seen, he concludes: "Though much has been lost, much remains" (93).

Does Abbey believe his own consoling words? Do we? My students don't. What compensates for the loss of the wild river? What is the cost of imposing on the earth the act of human will represented by Lake Powell? What kind of explanatory story is to be told, and in what language?

To me, "Down the River" is the core essay of all Abbey's nonfiction work. In it he writes both a paean and a lament for Glen Canyon as it was before the construction of the dam. Abbey marshals his creative and rhetorical resources to evoke a sense of what will be lost as the landscape is transformed:

> Down the river we drift in a kind of waking dream, gliding beneath the great curving cliffs with their tapestries of water stains, the golden alcoves, the hanging gardens, the seeps, the springs where no man will ever drink, the royal arches in high relief and the amphitheatres shaped like seashells. A sculptured landscape mostly bare of vegetation—earth in the nude.
>
> We try the walls for echo values . . . and the sounds that come back to us, far off and fading, are so strange and lovely, transmuted by distance, that we fall into silence, enchanted. (*Desert Solitaire* 144)
>
> [W]e launch off, in the middle of the afternoon, and paddle across the current to the shady side, abandoning ourselves once more to the noiseless effortless powerful slide of the Colorado through its burnished chute of stone. (*Desert Solitaire* 151)

The last clause creates a wonderfully complex, sensual image of simultaneous energy and lassitude. When hiking through the canyon of the Escalante, a tributary of the Colorado, agnostic Abbey reaches a moment of what can only be called religious ecstasy: "Under a wine-dark sky I walk through light reflected and re-reflected from the walls and floor of the canyon, a radiant Golden light that glows on rock and stream, sand and leaf in varied hues of amber, honey, whiskey—the light that never was is here, now, in the storm-sculptured gorge of the Escalante" (*Desert Solitaire* 154).

Decades later, in "Lake Powell by Houseboat," he tries to come to terms with what actually did happen to the place he loved. The language he finds and uses in his evaluation is curiously muted, revealing—in addition to his waning artistic energy—the pathos of a referential and Romantic vocabulary that has lost its object. It is almost as if now that the landscape that inspired some of his very best writing is buried under the silt of Lake Powell, a part of his linguistic power is drowned as well. But in his muffled grief and his groping for compensation, Abbey models for us one of the central experiences of modernity. For those who do not fully embrace modernization, Abbey's attempt to ward off despair is illuminating: we find a puzzling mix of honesty, self-deceit, repression, practicality, acceptance, bewilderment, and a sliver of hope and grace.

Someone who had read only the later essay might misjudge the intensity of feeling with which Abbey had loathed the idea of Lake Powell twenty-five years earlier. In the later piece he writes: "The lake itself, I'll confess, provokes a certain morbid curiosity. A part of my heart lies buried beneath that enormous pond" (*One Life* 86). A rather quiet statement. But two decades earlier in the 1960s, he can barely contain his bitterness:

> The time passes very slowly but not slowly enough. The canyon world becomes each hour more beautiful, the closer we come to its end. We think we have forgotten but we cannot forget—the knowledge is lodged like strontium in the marrow of our bones—that Glen Canyon has been condemned. We refuse to think about it. We dare not think about it for if we did we'd be eating our hearts, chewing our entrails, consuming ourselves in the fury of helpless rage. Of helpless outrage. (*Desert Solitaire* 164)

I look at the pictures of Glen Canyon available on the Web, and I share some of Abbey's emotion though all I see is a small, somewhat fuzzy set of pixels—about as far from the real quality of his experience as possible. Reading "Down the River" alongside "Lake Powell by Houseboat," we are moved to examine not only Abbey's but also our own ways of facing the bewildering mystery of modernity.

At the age of eight or nine, I discovered a volume called *The Book of Marvels*—an enchanted object. I sat for hours leafing through its pages, enthralled by black-and-white photographs of wonders from around the world. I can remember only three now. One was of Sugarloaf Mountain in Rio (which I swore I would one day see, and have not); the second was of the San Francisco–Oakland Bay bridge; and the third was an aerial photograph of Boulder Dam, as it was then called. I cleaved unto the picture of that dam as if it were a religious icon. And it was, for it instilled a feeling of the sublime. Something about the difference between the water level of Lake Mead on one side and the river far, far below on the other suggested enormous strength and power. I could sense the fearful weight of all that water pressing on the wall of the dam, and I did, in fact, marvel that human beings had constructed it. I had built dams of mud and sticks in the ravines near my home. I dreamed of backing the water up to the very top, but I was greatly satisfied with the foot-deep pond I occasionally managed. The feeling of power and control was unalloyed and intoxicating.

This was the middle of the 1950s, the high-water mark of "progress." I wonder now at my pure zeal for the great works of modernity. Outrage had a different meaning then for I felt it in response to a surprising list of targets. Why was it taking so long for them to finish the freeway into downtown Chicago? Why were all the settlements clinging unambitiously to the coast of Brazil? Why weren't people rushing to Brasilia? And what possible reason could there be for all that unbroken green area in the center of the map of South America? There were no roads there at all?! How could that be?! What was wrong with those people? Well, Americans should be living there—they'd fix things up right. And how come the Alaska Highway wasn't paved?

I was astonishingly impatient with the elephantine pace of modernity. There should be no place you can't drive. I'd look at the globe and fantasize a great bridge over the Bering Strait so that a smooth, unbroken road could stretch from Capetown to Cape Horn. When it was built I'd urge my parents to put us all in the new Pontiac V8 and take off. Looking back, I see my boyish impatience as exactly that—boyish. To my knowledge, the passion—that is what it was—for highways, bridges, dams, the whole genre of heavy construction, was not shared by girls.

I tell my classes in environmental literature about these fervent desires of my fifties childhood. They look at me quizzically as if they can't quite believe what they're hearing. And sure enough, today I am appalled at the North Cascades Highway—in my adopted state of Washington—which carved and diminished one of the nation's great wildernesses for no good reason that I can see (though that hasn't stopped me from using the road).

And I am amazed at the map of the 1960s plan for metropolitan Seattle with its grotesquely manic grid of freeways smothering the area and casually obliterating long-standing neighborhoods. Much of the plan was never implemented, but today a regrettable roaring freeway does stand on pylons in the midst of a beautiful Lake Washington bay and marsh. This freeway launches forth a couple of off-ramps to nowhere that serve as crudely obvious metaphors for the mentality that produced them. In the 1960s, the ramps were stopped in mid-air by an enlightened and rebellious vote of the people. They became the roads not taken.

But elsewhere in the city, an impossibly obtuse 1950s double-decker viaduct ruins what could have been one of the most glorious urban waterfronts in the world. As one promenades waterside, a tremendous cacophony emanates from the concrete monster looming almost directly above. It's like attending a classical concert while someone vacuums the aisles. Periodic cries are heard to tear down the structure, but Seattle these days can boast of the nation's second-worst traffic congestion, so the cries become whimpers and precipitate out with the winter drizzle. And I, unseemly as it may be, commute on that viaduct to work and enjoy the gorgeous vistas it affords the speeding driver of Puget Sound on one side and the city skyline on the other.

To be completely honest, alone late at night, wretchedly channel-surfing, I come across that cable show called something like "Big Machines," and my heart feels a little thrill of envy for the guy sitting atop the slag-hauler with the wheels the size of townhouses. Or for the guy operating the sublime monster mashing the guts of an open-pit mine into its maw. A female friend of mine caught me relishing this program, and a mixture of amusement, disbelief, and real alarm filled her face.

How many female civil engineers are there? How many whose specialty, whose passion, is dams? How much civility was involved in the building of the dam at Glen Canyon?

In the end, Abbey goes to Lake Powell. Barry Goldwater, of all people, said that the vote he regretted most in his life was his thumbs up to Glen Canyon Dam. The dam has not proven essential to either flood control or power generation. Abbey does do a nice job in his essay of describing its negative consequences: the lake is dangerous for smaller craft and is therefore a "playpen for the wealthy"; there are many fish but they are contaminated; the bathtub ring of the reservoir is called the Dead Zone, because that is in fact what it is; the shore is littered, and so on. But he can't generate the rage he felt when he first set out on the river long ago with the knowledge that it was doomed. Then he spat out these words:

> we float away on the river, leaving behind for a while all that we most heartily and joyfully detest. That's what the first taste of the wild does to a man, after having been too long penned up in the city. No wonder the Authorities are so anxious to smother the wilderness under asphalt and reservoirs. They know what they're doing; their lives depend on it, and all their rotten institutions. (*Desert Solitaire* 137)

In "Lake Powell by Houseboat," Abbey climbs up a side trail and manages to look out over thirty miles of wilderness that is still there. His juniper has survived and mule-deer prints are in the sand. He says that seeing these things "eased the guilt of houseboat travel." All this leads him to conclude: "There is much to be said for our American way of anarchy, after all. As civilizations come and go, rise, decline, and fall, ours is a pretty damn good one. Most things considered. While it lasts" (*One Life* 93). Quite a change in tone from his earlier anger. In the later essay, Abbey has chosen to focus more on what remains than on what has been lost. But the reader can't, because the power of the two essays is simply too unequal. One can agree, as I do, with Abbey's relatively favorable judgment of America but still feel that in the essay's context his affirmation is forced, his spirit defeated. Is it simply the mellowness of late middle age and literary success? Is it the fact that "Lake Powell by Houseboat" was written for the *New York Times Magazine* while *Desert Solitaire* was written in almost total obscurity? In the face of modernity, Abbey does not head for the hills with a rifle in his hands, like the hero does at the end of Louis Owens's novel *Wolfsong.* Nor does Abbey light out for the territory like Huck Finn—because where is the territory? It is now many hundreds of feet down under the water of Lake Powell. That's too deep to dive. A frightening depth.

Many years ago now, I set out on my first solo overnight hiking trip, my destination Fisher Lake, high in the wilderness of the Cascade range. Though the trailhead is only ninety minutes from Seattle and the trail only five miles long, when I reached the lake no one was there. It was mine alone. I was the last man on earth.

The night was a bit unsettling. I woke up frequently and thought I heard . . . I don't know what. A bear? A cougar? A sasquatch? But I finally fell asleep and slept soundly until morning. I relaxed. I felt quite peaceful and in harmony with the world. The sun glittered on the lake and began to warm the comfortable dirt of the campsite. But it turned out that the shimmering afternoon was to be the time of fearful visitation.

I sat on the shore after a sublime swim and was regretting that soon I would have to hit the trail and return to the city. The silence was pulsing gently in my ears. But I began to hear—and then feel—a rumble. It became more and more intense. Something big and violent seemed about to happen, but I didn't know what. My mind raced. With exponentially increasing alarm I thought: A landslide? An earthquake? A volcano? The sound actually became painful, and then instantly a jet fighter burst shockingly low over the ridge and shrieked across the lake like the sudden malevolent emanation that emerges in the midst of a normal dream and turns it into a nightmare. Everything on earth shook. In four or five seconds it was over.

At first it was as if I had jumped into water so cold that it stopped my breathing. Then, after a moment, I felt hollowed out. Unreal. Nothing was the same. Nothing could be trusted.

It was an instant, and no part of the landscape was destroyed or even altered. And though for me much remains, something had been lost. How much, I don't know, and don't really want to know.

I have to admit too that I wish I had never read "Lake Powell by Houseboat"; I wish that Abbey had not written it. But the ghost river of truth and time flows on, and cannot be dammed.

Note

Portions of this essay appeared in slightly altered form in *Organization and Environment*, 12, no. 3 (September 1999): 320–24.

Works Cited

Abbey, Edward. *Desert Solitaire*. Tucson: University of Arizona Press, 1988.
———. *The Journey Home*. New York: Dutton, 1977.
———. *One Life at a Time, Please*. New York: Henry Holt, 1988.
Lawrence, D. H. *Selected Literary Criticism*. Edited by Anthony Beal. New York: Viking, 1966.
Wordsworth, William. *The Fourteen-Book Prelude*. Edited by W. J. B. Owen. Ithaca, N.Y.: Cornell University Press, 1985.

THE BOYS' TRIP

RICK FAIRBANKS

"When is the boys' trip this summer?" It's my wife, Angela, setting the calendar for the summer. A staple of my life for the last fifteen years, the boys' trip is a kayak trip, usually on Lake Superior but sometimes on other northern lakes. It's more accurate to say that the boys' trip has been a staple of my life, in some form—canoeing or backpacking—since I was actually a boy, a ten-year-old boy to be exact, when I first went camping with a group of real boys (now that's a loaded phrase)—a canoe trip on the Missouri River. It became the "boys' trip" only a few years ago. I'm not quite sure who first called it the "boys' trip," Angela or Meg, the wife of one of the boys, my friend and colleague John. The "boys' trip" has always had a faintly mocking tone, at least in my ears, but despite this, perhaps because of this, it has become the "boys' trip" for me and, largely because of me, for the other boys as well.

The "boys' trip" is ironic, of course, most obviously because none of the "boys" are boys. I am the youngest at forty-eight. The oldest is sixty. A topic of conversation on every boys' trip is how much longer we will be able to keep going, how long until our backs won't let us sleep on the ground (not really the ground, but the ground mediated by a one-inch-thick self-inflating Thermarest pad), how long until our shoulders will ache so much at night that even liberal doses of scotch and ibuprofen won't carry us off to sleep, how long until it is all just more trouble than it's worth. I have to admit that the older we get, the stronger my sense of our ridiculousness grows. Maybe it will end when we are fully taken by the silliness of septuagenarians "camping out."

The "boys' trip" is also ironic because to one male friend at the place I teach I am "kayak man." To other colleagues, descriptions of this annual affair bring responses like, "oh, some male bonding." They must think of what they have read of Hemingway or Jim Harrison or Tom McGuane when they imagine what it is we do on these trips. I still vividly recall the night I revealed (a big mistake for a junior member of the department) that I actually still hunted, that I hunted and, occasionally, killed pheasants. Thank God I had enough sense to refrain from telling them that I had always found the killing part of hunting thrilling. After initial stunned silence they gently probed just how far gone I was. Did I actually like it? Didn't I think there was something atavistic about taking pleasure from killing? Again this fall, at our annual departmental retreat, when I regaled them with an account of a much too close encounter with a black bear (Angela insists it was an attack), I sensed a bemusement like that at my hunting revelation. "Why would anyone put himself in a circumstance to be charged by a bear?" Even as I write this a picture of a really big black bear hangs on the departmental bulletin board—one of my contributions to the depiction of last summer's highlights. The bear in the picture is about three times as big as the yearling that bluff-charged as I lay in my tent gazing—too directly, it would appear—at the bear some fifteen feet away. I'm sure I only strengthened their sense of me as Hemingway when, at our retreat where I had set up that selfsame, still unrepaired tent, I tried to get them to come down and look at the places where the bear had ripped open my tent with the swipes of its paws. No takers. I was disappointed, but then I'm the guy who half-jokingly complained to one friend that it wouldn't have been so bad if the bear had walloped me once—at least I might have an impressive scar. So, there I am: kayak man on the boys' trip. But then, why the "boys' trip?" Why have the boys claimed the description for themselves, even though we know it mocks us?

Well, first of all, we have the "boys and their toys" aspect of the trip. If you happened to catch us just at the moment when every last piece of gear and Ziploc bag of food was out of the vehicles but nothing had yet been put in the boats, you would find an astonishing array of stuff. The personal gear: sleeping bags, sleeping pad, inflatable pillows, headlamps, books (will two do for the week or should I throw in Nietzsche?), Tilley hats (owner's manual left at home), capilene shirts and underwear of varying weights (never cotton, "cotton kills"), "zip-offs," "Smart Wool" socks (merino wool, of course), rain gear, wind gear, paddling jacket (or maybe two or three for different temperatures and, of course, breathable), wetsuit, wetsuit booties, PFD (that's a personal flotation device, aka, a life jacket), paddle (for me a wooden paddle in a traditional Inuit style), spare paddle, paddle leash, paddle float, sunglasses, spare glasses, bug dope, sunscreen, spare batteries, global positioning system unit, maps, map cases, Crazy Creek chair (for comfort while drinking scotch, smoking cigars, and reading). The group gear: tents, stoves (two or three, requiring different fuels, for redundancy), water filters (again, multiples for redundancy, who can trust the water of Lake Nipigon or Lake Superior?), cooking pots, French coffee press, mini-espresso maker, utensil role, "Igloo" (a four-person emergency shelter, an antidote for hypothermia), one-person exposure bag (since Lake Superior is about forty-five degrees even in mid-July), towlines (yes, towlines), tarp, collapsible bucket, handheld VHF radio (usually more than one, for marine weather forecasts and for emergency transmission), flares, diemarker, booze (single malt scotch, of course, purchased at the border duty-free shop, but most likely also a couple of bottles of merlot or cabernet for the sun-dried tomato pasta dish or spaghetti), cigars (Dominican, Honduran—Will one a day be enough or should we plan for a layover, in which case, we'll need more?), food (no prepackaged, no dehydrated, usually a soft cooler full of fresh vegetables and dairy products).

You get the idea. As Emerson said, "Things are in the saddle and ride mankind." Now you know why kayaks are starting to turn up in ads for high-priced cars. A friend I paddled with for many years explained to anyone who asked that kayaking was about stuff, buying stuff. And the manufacturers know this. They know the market for stuff, gear—baby boomers who now have the money, if not the time, to go off in a kayak loaded with the very latest. There is a Darwinian pressure to diversify gear, to find niches into which new species of gear can flourish: "[T]he more diversified the descendants from any species become in structure, constitution, and habits, by so much will they be better enabled to seize on many and widely diversified places in the polity of nature, and so be able to increase in numbers." The kayak world has two kingdoms: touring and whitewater. The whitewater kingdom might as well be the mosses and clubworts. It's a blur of squirts, playboats, slaloms, what have you. In the kingdom of touring kayaks you will find recreational, general touring, ocean boats, and kids' boats. Each of these phyla can be further broken down by design (British, North American, traditional), material (fiberglass, Kevlar, plastic, "skin"—that is, painted canvas), and accessories (rudders, skegs, bilge pumps, day hatches). Evolution is occurring before our eyes. The pure British boat—skinny (no more than twenty-one inches wide) with tiny hatch openings, a cockpit barely big enough to crawl into, and heavy-duty fiberglass layup—appears to be on the brink of extinction. I own three kayaks: a P & H Baidarka Explorer—a pure British boat; a West Greenland "skin boat"; and an Aleutian skin triple baidarka. Both the Greenland boat and the Aleutian boat were built in a friend's kayak workshop in the traditional way—"skin" sewn on a wooden frame that is lashed together.

Some years ago I flirted with the idea that there was nothing more for me to buy. John wondered whether I would have reason to go on. With some relief I realized that it couldn't literally be true that I have nothing left to buy, since the evolution of stuff, of gear, presses relentlessly on in a Malthusian struggle for existence. There's always a low energy display

(LED) for the head lamp, a graphite version of a Greenland storm paddle, a handheld anemometer, a double-burner stove, a foot-operated bilge pump. And I dream of a new boat. I mean, I don't even have a day hatch, or a multiply-adjustable, slider-operated skeg, to say nothing of a deck recess for my spare paddle. My boat is fifteen years old. At kayak symposia (yes, kayak symposia), people react to my boat as they would to a whooping crane sighting: "I saw one of those on the roof of a car about five years ago." Last summer I was complimented just for paddling the old boat—like being noticed for driving a '57 Chevy to work.

So, yes, we are in it for the gear—the very word with connotations of armor, harnesses, tackle, implement, tools, even slang for "organs of generation" says the OED—and that makes it about right to call it the boys' trip. As much as we like to think of ourselves as different from the snowmobilers, bass fishermen, from the men who maintain a "cylinder count," I suppose we are consumers all the same. We shop. We look forward to the January issues of *Outside; Canoe and Kayak; Backpacker*; the buyer's guide issues. We buy *Sea Kayaker* for the reviews of paddling tops, sea anchors, parafoils, and dry bags. Although contemporary America displays an unparalleled genius for making us want gear, *this* image of the man in the wilderness, the well-appointed genteel man in the wilderness, isn't merely the construction of our consumerist age. It goes back at least into the nineteenth century. The British gentleman on safari, Teddy Roosevelt in Yosemite or on his North Dakota ranch, the Abercrombie and Fitch man, the Eddie Bauer man are each exemplars. In photos from his North Dakota ranching days, Teddy Roosevelt wore absurdly huge buffalo skin coats and enormous cowboy hats and studded vests and chaps. To be pictured as a cowboy, costumed as a cowboy, was to be a cowboy. My father, who never did anything more adventurous than camp out of a Falcon station wagon, excoriates the Eddie Bauer of the suburban mall: "Sir Edmund Hillary would throw up if he walked into one of those stores." Hillary was once the image of Eddie Bauer. The Eddie Bauer catalog of thirty years ago featured "expedition gear": mainly sleeping bags and coats containing only the "finest northern down." The message was that with the right gear you were already at the base camp on Everest. In my own backpacking days, the trip to the REI store in Seattle was as exhilarating as the hiking in the Cascades or on the Olympic coast. This Abercrombie image of man in the wilderness is one I am embarrassed to admit still has considerable force. If you doubt this, go to Canoecopia (!) in Madison, Wisconsin, some March or to the Canoe and Kayak Event in Minneapolis some April or any sportsmen's show that comes your way in the winter.

If we want to understand what our maleness has to do with wilderness, we had better pay attention to the allure of gear, to the power of gear to shape our experience. The right stuff indeed. We imagine ourselves posed in a picture in which our belonging, our knowing, is transparent. In an earlier day, the picture would have been of a group of men posed behind a huge stringer of fish or an array of ducks, geese, or upland game, but dressed in their cotton duck, their oiled-cloth, their shooting jackets, their Norm Thompson hats. In the early twenty-first century, we are posed in a mountain meadow but arrayed with our internal frame pack with its polycarbon fiber backsheet, in our Patagonia, our Tilley hats, our Danner boots, our Oakley shades. But this desire to be cool, to be outfitted, can't be all there is to the urge to spend time in the wilderness, because we could just stay home or direct the gear lust toward bicycles or fly rods. Gear might make the man, but the boys' trip isn't just about the gear.

We are told that masculinity is a construction, a social construction, and so too the "man in the wilderness" and the "manly pursuits." These claims are surely true, perhaps even obvious. However, usually you can hear a stage whisper "merely" or "just" before "construction," and the assumption is that, as mere constructions, these images of manliness will crumble to dust upon the discovery of constructedness. But this is not to take the metaphor of construction seriously enough. When we were kids my brother and some of his friends built a "clubhouse," a shack really, mainly because wood from the houses going up in our project was as plentiful as time in the late summer. We didn't want to spend any time in the clubhouse because it was so maniacally nailed together that the ceiling was a pincushion of protruding nails. The construction foreman, Lowell Carlson—the best friend any kid ever had because he would actually chase you all over the neighborhood after running you out of his houses—decided that the shack had to come down. He sent a couple of men over with crowbars to pull it down, but they gave up in frustration. Eventually they hooked a chain to it and dragged it in one piece to the landfill. "Constructed" doesn't mean ephemeral, unreal, or contingent, even if the construction, like the Abercrombie man, is one we would just as soon jettison.

Some philosopher, I forget who, compares our personal identities to towns that have lasted for thousands of years, an image originally applied to language in Wittgenstein's *Investigations*. At the core of the town might be the remains of a Roman settlement, perhaps surrounded by a walled crusader fort, around which are the factories of the Industrial Revolution. The image is one of Byzantine complexity—forgotten neighborhoods, streets that abruptly end, the new and the old jostling for space. Identities are surely constructed too, like the towns, out of elements that are too central, too integral to whom we are to simply be razed. Identities, selves, are like those towns, containing roads that dead-end at the walls, neighborhoods almost impossible to find, artifacts too precious to be razed. In modern hands an image like this is meant to explain why knowing oneself is no mean feat. Masculinity, maleness, and man-in-the-wilderness are complex things, like selves and old towns. They are comprised of associations, images, metaphors, analogies. As in personal identities and old towns, some elements in our ideas are too central to be razed and therefore constrain the ways concepts can be reconstructed. So, self-knowledge, understanding maleness, isn't urban renewal. We can't tear everything down and start over; we must find ourselves, our maleness, in a revision of the old town. I am reminded here of Michael Oakeshott's lyrical description of politics as "the pursuit, not of a dream, or of a general principle, but of an intimation" ("Political Education," 57). Politics, says Oakeshott, is the art of pursuing the intimations of a tradition. In politics we deceive ourselves if we think that pure concepts like justice or liberty are exportable as such into any tradition or culture. Like evolution itself, politics can work only with the materials that history provides. So it is with concepts. We can't start over because each of us is the product of a long and deep tradition of what it means to be a man. We will understand men, what it is to be a man, even a man in the wilderness, only by pursuing the intimations of these traditions. What I am looking for are those parts of the town that can't be ignored by the planners, the architects, the builders. As any architect knows, the new constructions force us to see the old constructions differently. The old parts of town come to have new meanings, but the old buildings won't allow us to build just anything.

As I think back over all these years of trips, one kind of memory remains most vivid, best exemplified by my first trip on the open waters of the Canadian north shore of Lake Superior. We are three or four days into the trip, the point at which you have settled into the pace and mind frame of traveling in a kayak. We are camped on Simpson Island or St. Ignace Island at the point at which Lake Superior finally turns east after tracking northeast all the way from Duluth. It is after supper, and I pick my way a couple hundred yards along the coast, crawling over boulders, jumping little inlets, as far as I can go before I run into cliffs that are ubiquitous in this part of the lake. What I didn't know when I set out is that I can see back what must be thirty miles, all the way back, it seems, to the Paps. The lake hasn't become perfectly glassy but is barely moving, desultory wavelets washing up seemingly on whim. In the shadows of the cliffs dusk seems nearer than it is. The rocks are giving up the heat of the day. I sit, not meditating, not thinking but in a numb awe at the immensity of the scene. Rousseau's account of reverie describes these moments best. "[E]ither lying in my boat as I let it drift with the water or seated on the banks of the tossing lake; or elsewhere, at the edge of a beautiful river or of a brook murmuring over pebbles," Rousseau experiences those instants about which he says, "I would like this instant to last forever" (*Reveries* 68). In these moments, he says we experience the sentiment of existence, "the precious sentiment of contentment and of peace" (69). It is an experience of "nothing if not ourselves and our own existence" (69), but—and this from a man who thought obsessively about himself and how others viewed him—the content of the experience seems to be nothing but the clouds floating in the sky, the waves lapping on the shore or against the boat, the rush of the water over the rocks and pebbles. In such moments our ownmost experience of ourselves, our experience of "nothing if not ourselves," is an experience of anything but ourselves. To experience the sentiment of one's existence is to fill one's consciousness with nothing but the sound of water moving.

I see now that my friend Dave is picking his way over the rocks, meaning to join me. When he makes it to the spot, he is apologetic. He realizes that he has interrupted the reverie. The talk is a little slow in coming, as if I were just waking up, but eventually the conversation turns to a history of his relationship with a long-absent father. This history will become familiar to me over the years, but this is the first time I am hearing it. It must be something about the spot, about my coming out of reverie, but I think he is telling me things he probably hasn't told anyone else. We talk until it is almost too dark to make our way back to the others.

Something like this scene has played itself out many times. The reverie isn't often as deep or complete. The conversation is not always so intense. The crucial elements are the sense of viewing something immense—the next twenty miles of Rainy Lake, the northern half of Lake Nipigon, across at Rock of Ages lighthouse fifteen miles way—the feel of nestling into a divot in the basalt or granite still warm from the day's heat, the talk of friends, stretching on until the cool of the lake overwhelms even the warmth of scotch and every stitch of clothing we have brought along. And when the talk is neither profound nor intense, the experience can be both: on Thompson Island, seeing the four visible moons of Jupiter; in the Slate Islands, wondering at the way the heat of the islands burned a hole through the fog that lay thick over the whole northern part of Lake Superior, watching, as dusk falls and the islands lose their heat, the fog retake the islands; once on the Minnesota north shore seeing the stream of light on the lake from a rising Jupiter; and, as far back as I can remember, on the Missouri River. I must be about twelve years old. The others have gone to bed. Ted Tarkow, our leader, then working on a Ph.D. in classics, and I. We sit at the fire, on a bluff above the river. We talk of stories, of myths—Tarkow, a maker of myths himself, always steered us toward myth and legend—of men like Lewis and Clark, like the Hidatsa and the Mandan, who sat on this river bank before us. And so always, the company of men, the easy fellowship of the boys, "seated on the banks of the tossing lake . . . at the edge of a beautiful river or a brook murmuring over pebbles" (Rousseau, *Reveries* 69).

I suppose my taste for the solitary is fed by the Romantic idea of the sublime, and specifically by my reading of Rousseau, Thoreau, and Muir. Though a powerful source of my experience of wilderness, a desire for solitude, by itself, isn't adequate to account for my sense of myself in the wilderness, because that experience is typically communal. The boys' trip is one of those rituals, like going out for a beer, talking in the steam room, hunting, playing in a handball tournament, without which most men are lost. Solo tripping is rare for me. The tradition of the sublime is inadequate to my experience for another reason. Rousseau's description of reverie, compelling as far as it goes, is for me too passive. Rousseau doesn't *do* anything, doesn't, dare I say, *master* anything. What I found compelling about Muir was his walking into the Sierra Nevadas with a loaf of bread to eat and an overcoat for shelter. I never had much of a taste for Muir's sense of the mountains as God's cathedral, although I might have had an epiphany or two over the years. But the idea of Muir up in the High Sierras for days or even weeks, living off his bread and whatever he could find to eat, curling up to sleep under an overhanging rock or in the roots of tree, now that was something. Now that was a man. I fantasized about emulating him by throwing away the backpack and the stove. It is well now to recall the old town, because we now find a Gothic cathedral next to a Roman bath. The Abercrombie man was defined by the stuff that made him prepared for anything. Muir is pure agency. All he needs to be prepared for anything are the bread and the overcoat. This too, or at least this aspiration, is part of the town.

Next to memories of reverie, what remains most vivid from the boys' trips are those occasions when I am prepared for anything, when I master myself in the moment. On one of those occasions Bruce Dickau, one of the boys, and I are leading a group of students along the Canadian north shore. We are paddling in the gap between Spar Island and Flour Island. The Lamb Island light station is off to the east, too far away, we decide, for an excursion. A light swell is running, left over from yesterday's two-meter waves—those rollers just big enough to be scary, making folks disappear in the troughs, but spread out enough to be manageable. I'm paddling close to shore. I know Bruce thinks I'm a goofball for doing this, since I occasionally scrape bottom, but I love the sensation of racing over the lake floor perfectly distinct a few feet beneath me. The big glacial erratics, car-sized boulders, seem to leap off the bottom as you paddle over them. I must have been in just the place where one of those leftover rollers hit a rock shelf or a big boulder lying on the lake floor, because I am jolted out of my paddling reverie by a steep wave standing up to eye level on my right. Without conscious reflection I raise the paddle to shoulder height, my hands just wide of my shoulders, the right hand not too far out of "the box" so that the force of the wave can't pull my shoulder right out of the socket, and I high brace by sticking the paddle into the wave face and leaning into the wave. It works perfectly; the force and energy of the wave make the wave, for a few seconds, into a solid thing. The high brace is as solid as hanging onto a dock or rolling up off the bow of a friend's kayak. I just hang on as the wave drives me to shore. When the wave breaks—did I shift to a low brace?—I sweep on the left and paddle like mad back into the lake. I get belted two more times before I make it out of the breaking waves. Everyone was sure that I would end up driven into the shore, but instead I manage to turn it into a lesson in the high brace, something I had practiced hundreds of times but never really had to do.

And this memory. Crossing from Michipicoten Island, some eleven miles off the Canadian shore where it turns north from the Pukaskwa area to run up to the Michipicoten River. The night before, the wind tried to blow us off the cobble beach. Three times I got out to pile more rocks on the corners of the tent. We are now maybe three miles into the crossing and thirty-five miles from the nearest human habitation. Flares and handheld VHF are worthless. No one is going to see us or hear us. I have never seen bigger waves, let alone paddled in them. They look like houses marching past me. Later, on the beach, Doug Olson says that he saw all three of us on a single wave face like rungs on a ladder. Phil Peterson, for years a racer and a sailor on Lake Superior, says they are ten feet, easy. The swells are running out of the southwest, all the way from Duluth, at least three hundred miles of fetch. The wind has backed to northwest and the breeze is still pretty stiff, so that we have wind waves, maybe a couple of feet high, running across the swells. I'm doing fine, not quite scared, at least I don't think so. I suppose I would have said that I was awed or amazed or numb at being out in those waves, but I am holding my own. And then I am upside down. I have no memory of how I got upside down; I have no idea what I thought hanging there. But, at least it seems so now, my body takes over and I roll up. Ever since I learned to roll, every time I got into a kayak, even in winter on the Mississippi, I would capsize and roll up at least once, just to know that I can roll in anything. But the only other time I had to roll—in later retellings I attribute the capsize to a hangover—I couldn't. I had to swim the boat to shore. But this time I roll up, and again, a second time in a three-hour crossing that will see me seasick, on the end of a towline, staring fixedly at what looks like a beach on the mainland. Anything not to look at the boys bobbing up and down on these huge waves, nothing but motion everywhere I look if not at that beach. Later, after the tension from the paddle has been released, we talk about the crossing. Doug says he's sure they could not have put me in back in the boat had I not been able to roll, had I been forced to wet exit. Deepwater rescues, techniques for assisting paddlers back into their boats, weren't designed for ten-foot swells.

So, yes, I want to be a master, to be able to rise to the occasion. This too is why I go into the wilderness. But this isn't mastery in the dreaded sense of domination. I am no master of nature but rather master of myself in the boat. I master the roll like one masters the cello. The description is a cliché, but the boat and I become one thing, an instrument. If mastery has to be mastery over something, then this mastery is over my middle-aged body. The skilled paddler doesn't master the waves but uses them, their force, for the brace and even for the roll. In the boat I feel liberated from my body. I'm fast, smooth, skilled. I can turn the boat just by edging it with a knee, lay on the water with a lazy sculling brace, exit the boat, climb back in upside down and roll it up full of water, roll a loaded boat anytime I want on a scorching hot day. I am able, skilled. This need to feel mastery also comes from somewhere. For as long as I can remember what I think I like most about the mountain men, the explorers, the cowboys that I have read about and watched were their devices, their skill in wresting not just survival but ease, comfort from the wild. This is Muir in the High Sierras but also the mountain man or the cowboy who lives from what he can carry on his horse.

If you want an image of the kind of man I'm talking about, you will find none better than John Wayne as Ethan Edwards in John Ford's *The Searchers*. From the opening image—a long shot from Martha's vantage on the porch of the Edwards home of something, at first we can't tell what, emerging, eventually emerging as a man, from the cactus and sage of the desert—to the last—the famous shot through the cabin door at Ethan walking away into the windswept desert as *we* close the door on him—you will find no more powerful image of the man who belongs in the wild. He doesn't talk, he does—he searches remorselessly for the captured girl, Debbie. He knows—his enemies the Comanche, the hard country in which they search. One of the great ironies of the film is how well he knows the Comanche—their language, their religion, their tools and weapons. We see him and Martin Pawley track the girl for five years, living, it seems, out of their saddlebags. He is able, skilled, but he is a twisted version of James Fenimore Cooper's Natty Bumppo. Edwards is a terrible man, a racist, a man twisted by hatred, a man whose search turns into a quest to murder the girl. That is the point, the genius of Ford's movie. He is *our* man. *This* was how the West was won. We who live beyond frontier have to close the door on Ethan Edwards because he is our man. We close the door in an act of pure repression. If we could, we would put him in a pine box, as Ford does Tom Doniphon, in his late, most elegiac western, *The Man Who Shot Liberty Valence.* But we can't really kill him. I suppose I should rather say that I, a baby boomer, raised on Westerns, raised in the West, can't kill him. How could John Wayne not be an image of manliness for a boy of the fifties and sixties?

Here, at least for me, is another essential part of the old town. This genre, the most dominant in the history of film, so central to American identity, although barely breathing today, is quite simply about the construction of maleness. As Jane Tompkins argues in *West of Everything,* the Western is about the ordeal of manliness, a man's struggle against death on the frontier. Tompkins's man of the West is scornful of words, a doer. He is devoid of an inner life, having been reduced to an element of the ubiquitous western landscape—Ethan Edwards emerging out of the desert, Shane emerging out of the mountains, Wyatt Earp riding out of Tombstone into Monument Valley. Tompkins sees the Western as a response to the popularity of women's literature in the nineteenth century. The Western novel and, later, film is a rejection of this kind of literature, of the East, of the novel of manners, of the Christianity so central to the women's literature of the nineteenth century. The genre continues on to Clint Eastwood in this mode of rejection, most vividly in the great Western *Unforgiven.* As perspicacious as Tompkins is, she misses something crucial—the Western is a continuation of a much older and deeper tradition. As both Jorge Luis Borges and the great film theorist André Bazin argued, the Western is the continuation of the epic. Male heroism has always been on the frontier, in the wild. Think of Achilles on the plains of Illium, of Aeneas leading the survivors of Troy into the west to their destined home, of Beowulf, of Roland on the frontiers of Christian Europe, even of Dante's pilgrim in hell and Milton's Satan traversing chaos to reach God's new creation. Any reasonably clever screenwriter could make the *Aeneid* into a Western—make Aeneas a wagon master, Dido the owner of a frontier saloon, the Laurentines, Sabines, and Tuscans, Comanche, Apaches, and Navaho. The wild, the frontier, has always been the space of heroism. The Western is anti-Christian because the epic is anti-Christian, essentially pagan, about the struggle for glory in the face of death. In the Christian era the epic is the primal scene of struggle between the pagan and the Christian.

The depth of this pagan tradition, the evidence that it still has a hold on us can easily be seen—witness the fascination with Shackleton, the *Lord of the Rings*, the seemingly insatiable need to invent frontiers (the first to ski across Antarctica, to climb Halfdome in a day, to ski down Everest, to solo to the North Pole, to descend the Bio-Bio), even Robert Bly's attempt to invent a myth of male heroism in *Iron John*. The boys' trip is the boys' trip because the frontier is the place where men become heroes, but we have no monsters to kill, no pilgrimages to complete, and no captured girls to find. We can do, we are able, only in attenuated ways. More often when I was younger, but sometimes still, I fantasize about a rescue that can only be performed in a kayak. But, by and large, our skill isn't *for* anything. The mountain climbers, the polar explorers, the first descenders of the world's rivers aren't heroes because they pursue only their own ends.

And Tompkins misses this too. She is right that the Western is an ordeal of masculinity, but she doesn't see that the ordeal, at least in the truly epic Westerns, is always put to the service of the community. The Ringo Kid, in Ford's first epic, *Stagecoach*, avenges the death of his brother, but he rids the town of the outlaws. So too Wyatt Earp in *My Darlin' Clementine*. Shane protects decent families from cattle barons. The story of Shane is virtually repeated in Clint Eastwood's latter-day Western *Pale Rider*, where Eastwood's Preacher protects settlers from rapacious miners who strip mine the land with a chemical bath. Glory is the price we pay for the hero's service. We sing the songs about him, tell the tales, and watch him ride into the sunset as payment for his sacrifice. We may rightly be squeamish about the ends for which the hero struggles—the defeat of Troy, the protection of Christian Europe, the taming of the West—but the heroes merely do what they are asked to do. The great Westerns depict the dilemma of the frontier hero: he does what we ask, but we send him away, close the door on him because civilization has no place for the savagery that makes him heroic. The ending of *Shane* is the classic expression of this trope. A man like Shane can't live with decent folk, he tells the boy. Even the *Aeneid* ends in a kind of cinematic isolation of the hero. The Western as a genre is elegiac because it is self-conscious that with the close of the frontier, we also close the door on the possibility of a certain kind of male heroism. It mourns that loss, especially when it contemplates the civilization that follows the close of the frontier.

This mourning—over the exchange of the lost frontier for the dubious gain of civilization—is felt most vividly in Ford's *The Man Who Shot Liberty Valence*. United States Senator Ransom Stoddard, played by Jimmy Stewart, returns to Shinbone, the scene of the heroism that made him

The Man Who Shot Liberty Valence. The film is structured as a flashback—Stoddard explains to the editor and reporter from the Shinbone newspaper who the man in the unmarked pine box is. In what critics say is either a stroke of genius or an old man's weariness, the black and white film is shot entirely on backlots and soundstages. Gone is the Monument Valley of *Stagecoach, My Darlin' Clementine, The Searchers*, and the cavalry trilogy. What we have instead are the flat, featureless, *movie* sets that represent Shinbone, but—and this is why we know that the film isn't the work of a weary old man—also the civilization that has been wrested from the West. The film, for anyone who loves Ford, is haunted by the absence of Monument Valley. The space of epic heroism has been replaced by tawdriness of Shinbone. This sense of diminishment is reinforced by the mise-en-scène of Stoddard's narration—a nondescript room dominated by a dusty stagecoach on blocks, the very stagecoach, it turns out, that Stoddard first rode to Shinbone. Given the elegiac tone of the film, we might also say that the stagecoach is the very one that carried the Ringo Kid—in John Wayne's first appearance in a Ford film—and six other passengers in the film that critics agree made the Western legitimate, Ford's 1939 *Stagecoach*. Stoddard, who as young man is the epitome of what Tompkins tells us the Western abhors—a man of law, of books, an easterner, a talker, a teacher, a man who waits tables and washes dishes, a "pilgrim," a "dude"—comes back to bury the quintessential westerner, John Wayne's Tom Doniphon. We learn that it was Doniphon who actually shot Valence. Standing unseen in the shadows (except by us, of course), Doniphon shoots Valence at just the moment Stoddard shoots at Valence, saving Stoddard's life and ridding the town, finally, of Valence. No one knows this save Stoddard, Doniphon, and his "boy," Pompey. Ford leaves us to wonder whether the senator's wife, Hallie, ever knows the truth. No one else will know the story until Stoddard tells the journalists and us. And while he tells the story, Doniphon lies in the next room in a wooden box, unknown to the journalists.

The film ends in profound sadness. The undertaker, from whom Stoddard demands Doniphon's boots, tells Stoddard that Doniphon is to be buried as a pauper. The editor is unwilling to tell the story, explaining: "This is the West, sir. When the legend becomes fact, print the legend." No one mourns Doniphon save Pompey. Ford's legendary West, the West of Doniphon, Ethan Edwards, Wyatt Earp, Captain Kirby York, Major Nathan Brittles, is dead. Doniphon doesn't even have a gun to bury with the boots. Stoddard sees his own hollowness and that of Shinbone and wonders whether it was worth it, whether the loss of the Doniphons, the very possibility of Doniphons, was a price too steep to pay for civilization of hollow men. Ford, at the end of an illustrious career, must wonder about his own hollowness, his own role in creating the legend of the "winning" of the West. The film's penultimate scene is shot inside the train car of the Washington-bound Stoddards. The scenery outside is nondescript, clearly projected on a soundstage screen. We can't help but think of the ending of *My Darlin' Clementine*—Wyatt Earp riding out of Tombstone into Monument Valley—or of *The Searchers,* with Ethan Edwards walking away from the Edwards' home into the desert. The frontier is closed and with it the epic space of heroism. The West is won. Hallie reminds us of this, saying to Stoddard: "Look at it. It was once a wilderness. Now it's a garden. Aren't you proud?" Stoddard doesn't answer but instead asks, "Who put the cactus rose on Tom's coffin?" The cactus rose symbolizes Doniphon, the desert, the West, the erstwhile romance between Doniphon and Hallie. "I did." The film closes with a final long shot of the train, heading east, the conductor's words, "nothing is too good for The Man Who Shot Liberty Valence," the last we hear.

Memory plays tricks, but this is the first film I remember seeing. I must have been six. My older brother took me. The Fairbanks boys' experience is no doubt at the core of my reaction to the film, and perhaps that accounts for its enduring power to haunt me, but haunted I am. Many of us of a certain age and especially those of us from the West or raised in the country, like Stoddard, like Ford himself, mourn the frontier and take that sense of loss to the wilderness—that thing, as William Cronon and others have shown us, that we civilized folk constructed after the frontier closed. The frontier was the space of heroism, and all we have is wilderness. The wilderness is haunted by the frontier. We are boys in the wilderness because we can't be men at the frontier.

But we aren't just haunted by the frontier hero. If we are to search for the intimations of our traditions, we have a rich resource in these heroes. If, as we have been told, masculinity's relation to nature is one of dominance, these heroes are exceptions to the rule. Even when they serve the interests of those who would transform, dominate, or master nature, as does Shane, they stand outside this domination, even mourn it. One can see this most clearly in Natty Bumppo, the ultimate American source of the Western hero, who can see the destruction of wild lands and therefore of his very way of being. We might use the frontier hero's alienation to purchase some irony about our own position in the most nature-dominating culture in the history of the world. Our experiences in the wilderness might lead us to ask the same kinds of questions about "civilization"—Ransom Stoddard's questions—that haunt the great Westerns. The western hero's silence and lack of interiority, observed by Tompkins, now come to have a different valence for those of us still drawn to wilderness. I seldom paddle solo, but I am always most drawn to it at the end of a school year, when I am quite simply sick of the sound of my own voice. The idea of some days of doing, of being consumed with the ordinary tasks of staying fed, warm, and sheltered, comes as a blessed relief from nine months of what feels like nothing but talk. The academic life is extreme, but many must feel the same pull of silence. The silence of a wilderness trip is an accomplishment, an antidote to the loquacity of our media-saturated lives.

Even when I paddle with the boys, I spend long hours in silent, almost meditative paddling. Sometimes the paddling, when the state of wind and waves doesn't require constant monitoring, can approach something like Rousseau's reverie—consciousness reduced to the sounds and sights of moving water. This is just what Tompkins sees in the Western hero as the objectification of interiority—the Western hero as part of the landscape. Oddly enough, Rousseau sees *this* experience as of "nothing if not ourselves and our own existence." That this claim would come from Rousseau, virtually the creator of modern subjectivity in *The Confessions*, is instructive, since Rousseau sometimes saw the human situation clearly out of the extremity of his own experience. It was Rousseau, as clearly paranoid as any autobiographer, who distinguished between the healthy self-regard, *amor de soi*, of the man of nature and the perverse, multiply-reflected self-regard, *amor propre*, of the civilized man. Who but Rousseau—who seems to be able to live only in the opinion of others—could see so clearly that the civilized man lives "always outside himself and knows how to live only in the opinions of others?" Who but Rousseau could see that civilized man "draws the sentiment of existence . . . from their [the others'] judgement alone" (*Discourse*, 81). Who but Rousseau, obsessed with authenticity, with telling the true story of the real self, could recognize that the deepest experience of the real self is to abandon oneself to consciousness of the natural world? Perhaps he simply knows better than anyone does what it is to be exhausted by the task of telling the true story of the real self in all its labyrinthine complexity. We would not want the complete reduction to the objective that Tompkins says characterizes the man of the West. Reveries like Rousseau's are powerful because they are exceptional. We don't want to become landscape, to become devoid of interiority, but in our age of self-help, self-love, self-knowledge, of obsession with self, it comes as a paradoxical blessing to hear that we might find ourselves in the full consciousness of what surrounds us, what endures without us. The experience of this discovery of self in the release from the self, surely both a Christian and a Buddhist paradox, is, at least for me, what epiphany in the wild means.

Here then is the deepest reason the boys' trip is the boys' trip—it is a return to the boy's mind, to the time before "Who am I?" "What do they think of me?" "What should I be?"—to simple summer days of just being and doing. Perhaps, as Rousseau would suggest, the boys' trip is a way to wrench the natural man out of the grip of the civilized man. But then it might be simplest to say that Rousseau's natural man is simply the boy. For me the first memory of this—of what Rousseau calls the sentiment of existence—is "Monkey Valley," my first empty place, just a valley with a stream that flowed in spring, a "hike" away from town. After eating our packed lunches, after exploring the stream, we would lie on a hilltop and watch the clouds float by. Then, again and again through the years, on the shore, at the summit, in camp, in canoes pushed along by the current, or kayaks pushed by the waves, over lunch, in the gathering darkness, with the boys, watching a hawk turn lazy circles, otters at play, pelicans turn slow and sinuous to land without making a ripple, waves break on the rocks, the fog drift in, a moose walk by in the eerie light of the full moon, the Northern Lights shoot up white and green, an osprey plunge down for its meal, an eagle swoop in to steal the trout from the osprey, a squall line approach across the water, the sun set huge and orange, the moon rise huge and luminous, the Summer Triangle emerge in the evening sky, Cygnus fly through the Milky Way, a distant lighthouse flash its barely discernable signal, tanker lights appear and disappear ghostlike, Jupiter rise out of the waters of Lake Superior, the wind fall down leaving the huge lake flat in the dying light of the evening.

Works Cited

Oakeshott, Michael. "Political Education." In *Rationalism in Politics and Other Essays*. Indianapolis: Liberty Press, 1962.
Rousseau, Jean-Jacques. *Discourse on the Origin of Inequality*. In *The Basic Political Writings*, translated by Donald A. Cress. Indianapolis: Hackett, 1987.
———.*The Reveries of the Solitary Walker*. Translated by Charles E. Butterworth. Indianapolis: Hackett, 1992.

"once a cowboy"

Will James, Waddie Mitchell, and the
Predicament of Riders Who Turn Writers

CHERYLL GLOTFELTY

I have never asked my dad why he wears a cowboy hat. I grew up in Palo Alto, California, better known today as Silicon Valley. My dad, Dr. Loren Acton, is a solar physicist who studies solar flares and sun spots. In 1985, he became an astronaut, flying aboard the Challenger space shuttle as a payload specialist. Dad has given more than four hundred invited talks to school groups, church groups, charities, university fundraisers, industrial seminars, and scientific meetings, in places as far flung as Sweden, Ireland, Gabon, Kazakhstan, the Czech Republic, Poland, Saudi Arabia, Italy, Japan, Costa Rica, Australia, and ships at sea. My dad generally arrives to these talks wearing a leather western sports coat, bolo tie, and cowboy hat.

Perhaps I never asked why Dad wears a cowboy hat because I think I already know. Dad grew up on the Airyland Stock Ranch outside of Lewistown, Montana, where his father raised registered Black Angus cattle. He rode a pony named Danger to a one-room schoolhouse in Fergus County and took part in recess pranks of letting the horses out of the corral so the kids would get excused from school to catch them. After a career in the California Bay area, my parents retired to Bozeman, Montana, where they built a house that overlooks the lovely Gallatin Valley. Dad glories in the crisp, clear air of the Big Sky country, drives a three-quarter-ton Chevy pickup, and likes nothing better than to spend a day exploring the bumpy back roads of Montana. "It was *just beautiful!*" he tells me. So I think that Dad wears a cowboy hat because he is proud that he comes from a ranching family, and because he loves the American West.

What is it like for a man to leave the ranch I wonder? My father's family moved to town when he was entering junior high, so he never had the chance to become a working cowboy. For him the decision to go to college and graduate school and to take a job in the city created no identity crisis. But what about fellows who worked on the land for much of their adulthood and then left? What is it like to *have been* a cowboy? Does such a man feel severed from nature? Does a cowboy feel less of a man when he loses his tan? Below, I follow the trails of two working cowboys who left the range to become writers, who then write about the range and their former lives as cowboys. In the one life, they earned their livelihood outside, in nature, working with animals; in the other, they labor indoors, representing nature, recalling animals. The cowboy is America's symbol of rugged masculinity, while the artist is often regarded as effeminate. What is it like for a man to lay down his rope and pick up a pen? To get off his horse and sit at a desk? To depict rather than to do the work of a cowboy? Why did these men decide to forgo riding for writing, and how has that decision affected their relationship to nature and their sense of themselves as men?

I trace the lives and works of two popular cowboy writers, Will James, from the first half of the twentieth century, and Waddie Mitchell, from the second half. Even by Will James's day in the 1920s, the open range cattle ranching that was widespread in the 1870s was thought to be a vanishing way of life, giving way to fenced pastures and feedlots. Yet both Will James and Waddie Mitchell found large free range outfits still afloat in the Sagebrush Ocean, principally in the vast high desert expanses of Nevada. Both men worked in and became a part of this country that others found so desolate. I'd like to ask them, "Gee, if you liked the cowboy life so much, why did you give it up? Do you miss it?" I'd like to know if they still think of themselves as cowboys.

The colorful character Will James is best known for his book *Smoky, the Cowhorse*, which won the Newbery Medal in 1927 and has never gone out of print. His twenty-four books include titles such as *Cowboys North and South; The Drifting Cowboy; Cow Country; Lone Cowboy; All in the Day's Riding; In the Saddle with Uncle Bill; Horses I've Known;* and *The American Cowboy* and comprise story and essay collections, novels, an autobiography, and children's books. All James's books are about aspects of cowpunching, including detailed descriptions of the people, places, animals, equipment, and work involved in cattle operations throughout the West. As he claims in many a preface, all his books are drawn from his own experience as a drifting cowboy, bronc buster, cattle rustler, rodeo contestant, Hollywood stunt man, and all-around cowpuncher.

Although his books conceal his true origins behind a mythical story of being born on an emigrant trail in Montana to a longtime ranching family and orphaned by the age of four, biographer Anthony Amaral discovered a quarter century after his death that Will James was born Joseph-Ernest-Nephtali Dufault in 1892 in Quebec, Canada, one of six children of a French-speaking merchant family (Amaral 101).[1] In 1907, at the age of fifteen, Ernest bid farewell to his family and hopped on a train bound for the West, where he worked odd jobs, looking for a way to ride horses and become a real cowboy. By 1910, Ernest had made his way to Montana and ditched his French name, eventually adopting the American-sounding Will James. Will drifted around the West, zigzagging from Canada to Mexico and back again, always drawn to unfenced territory, open range, where, as he says in *Cowboys North and South*, "I can spread my loop without getting it caught in a fence-post" (63). Wherever he went, Will James left a paper trail of sketches of cowboy scenes, which he gave to friends and employers and left tacked to the walls of ranch houses, barns, and barrooms.

As early as his mid-twenties, James apparently contemplated quitting cowboy life in order to become an artist. While an inmate of the Nevada State Prison in Carson City in 1916, where he was serving time for cattle rustling, James made a three-part sketch entitled "The Turning Point," which depicts the "Past" with himself roping a Texas longhorn; the "Present" with himself sitting behind prison bars, thinking; and the "Future" with himself standing before an easel, painting a picture of a cowboy on horseback. The caption reads, "Have had ample time for serious thought and it is my ambition to follow up on my art" (Amaral 82ff.). Nevertheless, after his release from jail, James returned to drifting and range work until "providence played a hand," and in 1919, while fooling around with some friends, James was bucked off a horse named Happy in Reno, Nevada, smashing his head on a train track, after which the horse stepped on his back. Recalling that fateful ride in his autobiography, James observes: "That last bronk I'd rode had sure fixed me so I'd have to be an artist. . . . But how to be an artist was what had me stumped" (*Lone Cowboy* 417).

As luck would have it, James met a man who offered to write him a letter of introduction to the editor of *Sunset* magazine. In January 1920, after some initial rebuffs, *Sunset* ran a full-page drawing by James and thereafter purchased and published many more. Flush with good fortune, James married sixteen-year-old Alice Conradt, but by the end of 1920, *Sunset* informed James that they had a backlog of his drawings and didn't need any more. Frustrated by a series of rejections from other magazines, James took Alice's suggestion to write a story to accompany his drawings. *Scribner's* accepted his very first effort, "Bucking Horses and Bucking-Horse Riders," which appeared in 1923, kicking off a successful career of writing illustrated stories and books about cowboy life for the entertainment and education of urban audiences. James reflects: "'Happy' had proved to be my turning point, a mighty rough one maybe, but my turning point sure enough. . . . most likely if I hadn't met up with 'Happy' . . . I'd still be riding for a living—not that I'd mind that so much, but this ain't so bad either, and being a has-been I'd rather draw and write about that life than take a back seat as a rider" (*All in the Day's Riding* 85).

"The Turning Point" by Will James
(COURTESY OF SPECIAL COLLECTIONS, UNIVERSITY OF NEVADA, RENO, LIBRARY)

As author and illustrator, James's hallmark was to paint the working cowboy as he really was, countering the distorted six-gun heroics presented in popular Westerns. In the preface to his first book, *Cowboys North and South*, James names his credentials as having lived the life he writes about, conveying that life in the language actually spoken on the range. Above all he prides himself on his authenticity: "I was born and raised in the cow country, I am a cowboy, and what's put down in these pages *is not material that I've hunted up*, it's what I've lived, seen, and went thru before I ever had any idea that my writing and sketches would ever appear before the public" (x). Both James's subject and his rough-and-ready style exuded masculinity. One contemporary reviewer observed of James's autobiography: "Here is no lilyfingered, typewriter-tapping dude with the background of a two-week visit to Santa Fe" (quoted in Amaral xiii).

James was particularly drawn to large expanses of unfenced territory where both cattle and cowboys roam free and where the seasoned cowhand will "get to feel that you're a part of that desert, that you belong to it" (*The Drifting Cowboy* 91). Despite its hardships, cowboying is depicted as the perfect job, attuning a man to the land while preserving his freedom. In contrast, "easy" work in town cuts a man off from the land, ties him down, and causes him to weaken. In "Once a Cowboy," James tells the story of a cowboy who moved to town to escape the hardships of bad weather and to secure his financial future. After about one month of working indoors at the Feed Market, the cowboy notices that he is losing his appetite, "And what's more, my complexion was getting light, too light" (*The Drifting Cowboy* 22). When spring arrives, the narrator observes: "Everybody I'd see would remark how great it was outside in the spring air, and rubbing their hands would get to work at the desk and typewriter, and forget all about it the minute they set down. I felt sorry for 'em in a way, 'cause it struck me as though they'd never had a chance to really appreciate springtime—or was it that their years in captivity that way had learnt 'em better than to hanker for such?" (24–25). The story ends happily with the cowboy reclaiming his stabled horse and riding back out to the range. The irony, of course, is that it took James at a desk to create this story about a cowboy who rejects captivity.

While James came to earn his living indoors at a desk, he continued for all his life to find solace in big stretches of open country, and his sense of himself as a man remained tied to his accomplishments and reputation as a rider. If one goes strictly by what James wrote in his books, his new vocation allowed him to enjoy the best of both worlds. The huge success of *Smoky, the Cowhorse* enabled him to buy a large ranch in Montana, which he named the Rocking R. He writes contentedly at the conclusion of *Lone Cowboy*: "Now—I've finally gathered me a little scope of range like I've always hankered for—A place away from lanes, and in the heart of a wide-open cow and horse country—only a hundred miles from where I was born—I have my ponies, cattle, corrals and all to my taste—There's hundreds of wild horses around, thousands of cattle from neighboring outfits—timber—big creeks with trout in 'em—plenty of grass on both sides . . . —I'm at home—" (433). In a letter written several years later, James observes to a friend that "Mixing my writing with my riding is my idea of fine and peaceful living" (quoted in Amaral 148).

Self-portrait of Will James at his writing desk
(COURTESY OF THE WILL JAMES ART COMPANY)

James depicts himself as having achieved a satisfying synthesis as a cowboy writer on his ranch. Self-portraits picture him engrossed at his writing desk, wearing a cowboy hat and cowboy boots with spurs. But the halcyon image of a contented cowboy writer and artist at home on his ranch may be as spurious as the tale of his origins. In fact, as Amaral documents, James was on the lecture circuit or in Hollywood much of the time to earn money for mounting ranch expenses and personal bills. He abused alcohol and mistreated Alice, who eventually filed for a legal separation. In one of his alcoholic sprees, James sold his ranch for a mere one thousand dollars, and he died in 1942 at the age of fifty from alcohol-related complications. One acquaintance speculated that James became embittered at having to give up riding at such a young age, while Amaral surmises that James was eroded and ultimately destroyed by guilt about denying his true family. According to Amaral, James visited his family in Canada for the last time in 1934 and to his mother's dismay insisted that all his letters, telegrams, pictures, photographs, and anything else linking him to Ernest Default be burned. He wrote his brother a letter, confessing that "I often wish that I hadn't misrepresented myself as I did, but I couldn't dream of the success I've had and now it's too late to change. And if what you all know ever got in the right hands I'm so well known now that it would be in all the papers overnight and I'll be classed as an imposter which would ruin me for good. Of course I'm not an imposter but I'd be classed as one, and after that I'd just as well go and bury myself" (quoted in Dudar 12).

As an author, James walked a fine line between fidelity and embroidery, tall tales being characteristic of the western storytelling tradition in which he participated. James prided himself and capitalized on his firsthand knowledge, a factual grounding that popular western novelists of the period often lacked. But to win an audience, James also had to be an engaging storyteller, slipping in entertaining confabulations that, however, must never venture beyond the plausible. According to the western code, a man is as good as his word. The worst insult one can level at a cowboy is that he's a liar. James worried that if the fiction of his origins were discovered and revealed, it would discredit everything else he wrote and destroy his reputation. As a successful writer, James paid a high price for lying and for his youthful misconception that real cowboys don't grow up in Montreal.

Being the real thing is also a selling point for contemporary cowboy poet and entertainer Waddie Mitchell, of whom newspaper columnist Cory Farley writes, "Maybe what gives Waddie Mitchell his authenticity is his authenticity." Bruce Douglas Mitchell, nicknamed Waddie by his father (Waddie is an old term for cowboy), was born in 1950 on a remote ranch in northeastern Nevada. The young Waddie enjoyed listening to the talk of working cowboys, who entertained one another by telling stories, sometimes put to rhyme. Back then Waddie never thought of these oral recitations as poetry: "No, poetry's for sissies and womens," he used to think (*Waddie's Whole Load* audiocassette). Nevertheless, he memorized his first cowboy poem at the age of five and wrote his first poem in about 1973 (Mitchell, interview). At age sixteen, Waddie quit high school to become a cowboy, a calling he pursued for the next twenty-six years, eventually managing ranches so remote that he was lucky if he made it into town once a month ("In the Spotlight"). He married a woman—named "Toot" in his poems—who also enjoyed ranch living, and they had five children.

Life began to change for Waddie in the early 1980s when a PBS film crew featured him in a documentary about the last working cowboys entitled *The Vanishing Breed* (Brennan). T. V. host Johnny Carson became aware of Waddie and invited him to be a guest on the *Tonight Show*. As the story goes, because Waddie could not be reached, Johnny Carson's crew telephoned a neighbor who drove forty miles to inform Waddie of the invitation. The busy buckaroo initially declined because it was calving season, and besides he had never heard of Johnny Carson ("Waddie Mitchell: The Cowboy Poet"). Nevertheless, he did appear, making such a hit that he was invited back on the show several more times. In 1984, Waddie helped his friend Hal Cannon organize the first annual Cowboy Poetry Gathering, held in Elko, Nevada, in January 1985.[2] Over the years, Waddie won a large following at the Elko Cowboy Poetry Gatherings, and he was signed on by Warner Brothers to make one of the first recordings under its new Warner Western label, which resulted in the 1992 album *Lone Driftin' Rider*. It was at this time that it looked like Mitchell might be able to succeed as a professional entertainer, and he was faced with the tough decision of whether to forgo ranch work for a new career as a touring cowboy poet/entertainer.

Earlier poems provide oblique hints that Mitchell had been getting restless and was perhaps ready for a change. In a Christmas poem entitled "There's Nothin' Like Nothin'," collected in his first book, *Waddie Mitchell's Christmas Poems* (1988), Waddie regrets that after fourteen years of married life on cattle ranches, he "didn't have much to show / For the life he'd spent ridin' for cattle / And he was feeling especially low." This poor cowboy doesn't even have enough money to buy his wife a Christmas present. He asks her, "Do you think I am wrong punchin' cattle? / And should I find me a good job in town?" but she reassures him that she likes the life they've chosen and that "There's nothin' like nothin' for Christmas, / When I know it's comin' from you" (32–35). In a later poem, "Haven't Sold Your Saddle," composed when Waddie was approaching forty, the narrator complains to his friend that "My wife has started gainin' weight and gray shows in her hair; / Her existence seems to be in runnin' kids from here to there. / My job has lost its challenge; it seems like it never changes. / Sometimes I'd like to chuck it all and leave to ride new ranges" (*Waddie's Whole Load* 82). In "Morning Soliloquy," a cowboy in winter suffers from the cold, low rations, a toothache and a sore knee, a pack rat whose ruckus disturbed his sleep, and troubling "dreams of the youth I squandered." He wonders to himself, "I don't know why I still do it; / Ain't easy on these outfits alone; / But it ain't like I've got many choices— / This lifestyle's all that I've known" (*Waddie's Whole Load* 76).

Waddie's skillful delivery of poems such as these soon opened up choices for him. His poem "Where to Go" recounts this difficult crossroads 'twixt roping and reciting. A young cowboy confides to his friend that he's having a hard time figuring everything out, "some think I've got something special, / And to go for it's all that's essential, / If I'd give it my time and my effort, / With hard work I might reach my potential." Others tell him that "I'm flyin' too high for my own good / And it's time I come back to the ground." The friend counsels him to saddle up and go off by himself, "Find sanctuary out on the cow range. / Let the wind do its thing on your mind. / . . . Trot off 'cross the desert and search for that trail / That will help you find out who you are." The young cowboy returns with a clear sense of direction, "He would sow and reap his own lot" (*Waddie's Whole Load* 28–29).

Since that fateful decision, Waddie's career as an entertainer and cowboy poet has taken off. He spends approximately three hundred days a year on the road, earning between $2,500 and $10,000 per appearance (Crowley). He has performed for audiences around the world, from Melbourne, Australia, to Zurich, Switzerland, and has appeared on television for the *Tonight Show, Larry King Live, Good Morning America,* TNN, the History Channel, PBS, CMT, and BBC. He has also been featured in *People, Life, USA Today, Fortune, National Geographic,* the official program for Super Bowl XXX, and the *Wall Street Journal.* He has recorded numerous albums for Warner Brothers and the Western Jubilee Recording Company; starred in documentaries about his life and the cowboy way; published a second book, entitled *Waddie's Whole Load* (1994); and has been inducted to the Cowboy Poets and Singers' Hall of Fame. Waddie says that his goal is to one day buy his own ranch: "I'm hoping for the opportunity to go broke on a ranch by myself instead of helping somebody else do it!" ("Waddie Mitchell: The Cowboy Poet").

Has Waddie been happy with the choice to leave buckarooing for entertaining and writing? As is true of most of life's decisions, the results are a mixed herd. On the dark side, homesickness sometimes hits hard when he's on the road. In a poem entitled "Puttin' Things Right," the narrator likens Nevada to a woman, "And I miss her so this evening— / I'm obliged to be away. / We often don't know what we have / 'Til once we start to stray. . . . / Her cow range beckons me to come / And ride it once again. / Her mountains lure my soul back / Where I communed with Him" (*Waddie's Whole Load* 18–19). He misses all the things he loved most about buckarooing, including work that put him in daily, intimate contact with nature. A celebratory poem entitled "Commutin'" that Waddie wrote before he left cowpunching becomes nostalgic when he declaims it as a professional entertainer:

There ain't nothin' like the feelin'
Thet ya git down deep inside
As ya trot out in the mornin'
When you've hired on to ride

And your mount's enthusiastic
And the air is crisp and new
And there's lively conversation
Goin' on amongst the crew.

There's some bridle crickets chirpin',
Jingle bobs tap out a tune;
On one side the sun is risin',
Just ahead there sets the moon.

> Shadows high trot there beside ya,
> Elongated, keepin' pace,
> Reassurin' you ain't hobbled
> By restrictive time or space.
>
> Out in front the boss is postin'
> To the same beat as his song,
> And the realization hits ya
> That you're right where you belong.
>   (*Waddie's Whole Load* 49)

In contrast to that life of freedom on the range, Waddie's new profession takes him to cities, where people "choke in crazy traffic jams, / Fight for seats on bus or train. / It's a wonder that this ritual / Doesn't drive them all insane" (*Waddie's Whole Load* 50). Another rocky place in Waddie's new path has been the break-up of his marriage, the pain of which is expressed in his poem "So Darn Hard": "It happens all too often in this current day and age, / Through mistakes and poor decisions that are made along the way, / From a small misunderstanding, soon a full-scale war's engaged, / And we find one day that . . . we're alone. . . . / But it's so darn hard to be alone, / To be out there in this cold world on your own, / To be apart from everything that is comf'terble and known" (*Waddie's Whole Load* 34).

On the bright side of his midlife change in mounts, Waddie has indeed been able to reach his potential, to realize the full flowering of his special gift for storytelling and verse. Waddie has said that taking his poetry on the road has given him "a brand new lease on life" and broadened his horizons ("Featured"). Formerly isolated on remote ranches, Waddie now enjoys the opportunity to travel the wide world and to meet the larger human community. "For twenty-six years, I loved my cowboy'n life," he told an interviewer, "I did it professionally. Now, I have the opportunity to get around the world and I know that there's a lot of good people out there whose lives are enhanced by the spirit and the values of the 'Cowboy'" ("Featured"). Needless to say, Waddie's entertainment career has earned him a gunnysack full of money, providing financial security for his family, and increased choices for his future.

Affluence has also allowed Waddie to indulge his taste for fine togs. According to an article in the *Saturday Evening Post*, Waddie, "one of the dandiest and best liked of all storytellers, . . . dresses in $600 boots with his brand imprinted on them, a handmade horsehair belt with a gold-inlaid, sterling silver buckle, and a beaver-felt black hat" (Crowley 1). His fantastically long and impeccably waxed handlebar mustache is legendary, and he has even written a poem about it. As a performer, Waddie teases the audience's preconceptions of masculinity as he plays his prettied-up appearance—more costume than workingman's clothing—against his seasoned, around-the-campfire tone and cowpoke subject matter. For insiders, such as the ranchers who attend the annual Elko Cowboy Poetry Gatherings, Waddie is beloved for his public defense of the cowboy way of life and for his natural gift of entertaining. He is one of them, and they fondly remember the Waddie they knew before he became famous, and they like him because despite his fame he has not taken on airs and will still have a drink with folks at the bar after the show. For urban audiences, such as the full house at Carnegie Hall in New York City earlier this year, Waddie evokes a spirit of the Old West when men did outdoor work, bedded down by the chuck wagon, and rode on trail drives. Even if one has never even remotely lived this life, it is part of an American heritage that most audiences embrace as their own. Waddie reassures his audiences that this is a *living* heritage and that there are still working cowboys performing their seasonal rounds, depending on public policies that will allow them to continue.

What is it like for a man to leave the ranch? As my dad might say, "'t'ain't easy." Nor is it easy to stay. In the case of both Will James and Waddie Mitchell, leaving ranch life coincided with the onset of middle age, in James's case an early middle age due to an exceptionally battering young manhood. The question of leaving the ranch, then, becomes intertwined with the more universal question of what is it like for a man to lose his youth, or, to put it more positively, what is it like to grow up, to grow away, to grow into a new phase of life? As both James and Mitchell suggest, for some young men there is nothing quite so appealing as being a cowboy. Living a life outdoors with animals, working alongside experienced hands, learning the craft, riding a horse, enjoying the routine of the trail, living on the western range—there is just nothing to compare in school or in town life. It is an ideal life, the perfect way to become a man. One's horizon is not complicated with women. It is free. One is unencumbered.

But there comes a time for most men when they want to get married and possibly have a family. The pull of biology is powerful, the attraction of women is great, and the single life becomes almost unbearable. Even the most supportive wives, who love ranch life themselves, cause these cowboys to think differently about their chosen job. There is never enough money. A man's sense of self-worth is eroded when he cannot adequately support his family. He wants to own a house and to have his own ranch. But the opportunity in cowboying to get ahead is minimal, and so he gets frustrated, exacerbating the normal irritations of this life, such as working on Christmas. Cowboying starts to seem routine and dead-ended. The cowboy life being no longer new and novel, he casts about for a way up the financial ladder. He has aspirations, ambitions. Added to this, if he has been injured, cowboying gets too rough on the body. A man wants comfort, an easier job physically. But some cowboys know nothing but cowboying. Some have not gone to school, know nothing about city work, can't stand the thought of store clerking, and can think of nothing else. Both James and Mitchell were lucky in that they had nascent talents that they could develop and eventually market. Unlike many of their peers, they did have a way out, and they took it.

Here begins an exciting midlife period of developing their artistic talents and realizing their new potential. They strive, work hard, learn, struggle, pursue opportunities, meet new people. This is an exciting time, a new lease on life, full of promising new prospects and adventures. Freshly off the range, they find their cowboy experience to be valuable capital for their art. There is a market for it. It is exhilarating to capture on paper or in performance every wrinkle, nuance, mood, and memory of cowboying life, to document one's earlier life and find an eager audience for this material. For the first time, the former cowboy begins to make big money, and this is very good.

After the artistic life itself becomes routine, however, the question mark looms. Will James did well for nearly two decades of drawing and writing professionally but ultimately could not keep up the pace and fell prey to alcoholism. Shattered relationships, mounting regrets, compounding expenses, mismanaged money, and mishandled friendships haunted his days, and his life crashed down around him.

After one decade as a professional cowboy poet, Waddie Mitchell's future is still to be written. It seems to me that Waddie is using the new chapter of his life as an opportunity for growth. He may not be as happy as a young cowboy, but I bet he's happier than a cowboy in a midlife crisis. Waddie is riding this horse of fame as a cowboy poet for all it's worth right now, but who knows how long he can stay on? Like Will James before him, Waddie has emerged as an ambassador, publicist, and defender of the cowboy way of life. For James, the principal human threat to open range ranching was "nesters," who cut up the land with fences. In our own day, the threats confronting the cowboy have multiplied like flies on a cow pie. Waddie laments in "The Throw-Back" that "nowadays they've throw'd us some ringers, / New problems that's kicked in our slats, / Like computers, the futures and unions, / And worst of all—bureaucrats" (*Waddie's Whole Load* 101). For Waddie, being a cowboy poet—while ending his own cowboying life—has become an effective way to celebrate and perpetuate cowboy culture in the face of forces that would cause its demise.

CHERYLL GLOTFELTY

The cowboy today faces an uncertain future. Is he a vanishing breed or is he here to stay? In either case, most Americans will have little or no contact with working cowboys, and the image of the cowboy will continue to be distorted on the silver screen. What we learn of real cowboys and ranching culture will be thanks to writers and performers such as Will James and Waddie Mitchell. As we have seen, their experiences illuminate some of the pleasures and perils of riding/writing the range between two cultures, rural and urban, cowboy and commuter. Both men write of, from, and for cowboy culture and work life, using the cowboy idiom, and marking themselves by clothing and speech as cowboys. For both men, to an extent, the personality has become the product, as they are not only authors but celebrities. Consequently, both men have been put in the ironic position of performing themselves, of playing themselves as cowboys in documentaries and public appearances, even after they have ceased being cowboys in the sense Waddie Mitchell had in mind when he told his son that "cowboy's a verb, not a noun: / It's what you do more than a name" (*Waddie's Whole Load* 85). Yet, although they can no longer claim to be cowboy-as-verb, both Will James and Waddie Mitchell continue to think of themselves as cowboy-as-noun. Even as they tour the world and mingle with sophisticates, both men dream of owning a ranch of their own, where they can once more be in touch with nature, with cowboys, with animals, and, just maybe, with their most cherished but elusive identities as free men of the open range.

Notes

For their generous help, I would like to thank my father, Dr. Loren Acton, of Bozeman, Montana; Dr. Barney Nelson of Sul Ross University, Alpine, Texas; and archivist Steve Green at the Western Folklife Center in Elko, Nevada. Grateful acknowledgment is made to Waddie Mitchell for permission to reprint the poem "Commutin'."

1. For an interesting study of orphan imagery in cowboy culture, including a discussion of Will James, see Allmendinger's chapter "Where Seldom Is Heard a Discouraging Word: Orphanhood and Orality at Home on the Range," in his *The Cowboy: Representations of Labor in an American Work Culture*.

2. Stanley and Thatcher's fine scholarly anthology *Cowboy Poets and Poetry* provides a good history of the Elko Cowboy Poetry Gathering, also including treatment of cowboy poetry and poets.

Works Cited

Allmendinger, Blake. *The Cowboy: Representations of Labor in an American Work Culture*. New York: Oxford University Press, 1992.

Amaral, Anthony. *Will James: The Last Cowboy Legend*. 1967. Reno: University of Nevada Press, 1993.

Brennan, Sandra. "Waddie Mitchell: Biography." Linked to www.artistdirect.com/music/artist/bio/ (accessed 29 June 2002).

Buckaroo Bard. Directed by David Broberg, John Clemons, and Dick Jamison. Performed by Waddie Mitchell, Richard Farnsworth. Salt Lake City: Brigham Young University, 1988.

Crowley, Carolyn Hughes. "Ropin' and Rhymin'." *Saturday Evening Post* 267 (January–February 1995): 74–76. Linked to http://firstsearch.oclc.org (accessed 29 June 2002): 1–3.

Dudar, Judy. "The West *d'un Cow-boy Solitaire:* Will James." In *Literature of Region and Nation: Proceedings of the Sixth International Literature of Region and Nature Conference,* edited by Winnifred M. Bogaards, 8–16. New Brunswick: Social Sciences and Humanities Research Council of Canada, with University of New Brunswick in Saint John, 1998.

Farley, Cory. "Waddie Mitchell Honored As One of the Top 20 Nevada Artists, Authors & Entertainers of the 20th Century." *Reno Gazette-Journal,* 21 November 1999. Linked to www.somagency.com/WaddieMitchell/ (accessed 29 June 2002).

"Featured at the Bar-D Ranch." www.cowboypoetry.com/waddie.htm (accessed 29 June 2002).

"In the Spotlight: Waddie Mitchell, Cowboy Poet." *iBerkshires.com.* Linked to www.iberkshires.com/qa/ (accessed 29 June 2002).

James, Will. *All in the Day's Riding.* 1933. Cleveland: World Publishing, 1945.

———. *Cowboys North and South.* 1924. Missoula, Mont.: Mountain Press Publishing, 1995.

———. *Cow Country.* New York: Grosset and Dunlap, 1927.

———. *The Drifting Cowboy.* 1925. Missoula, Mont.: Mountain Press Publishing, 1995.

———. *Lone Cowboy: My Life Story.* New York: Charles Scribner's Sons, 1930.

Mitchell, Waddie. Interview by Deb Spring. "Home Means Nevada: Folklife in the Silver State." KOLO radio. Reno: Nevada State Council on the Arts, Folk Arts Program, 1986.

———. *Waddie Mitchell's Christmas Poems: A Cowboy Celebrates Christmas.* Salt Lake City: Gibbs Smith, 1987.

———. *Waddie's Whole Load: The Cowboy Poetry of Waddie Mitchell.* With audiocassette, "A Chat with Waddie." Salt Lake City: Gibbs Smith, 1994.

Stanley, David, and Elaine Thatcher, eds. *Cowboy Poets and Cowboy Poetry.* Urbana: University of Illinois Press, 2000.

"Waddie Mitchell." www.wbr.com/nashville/warnerwestern/cmp/waddie.html (accessed 29 June 2002).

"Waddie Mitchell: The Cowboy Poet." Scott O'Malley & Associates. www.somagency.com/WaddieMitchell/ (accessed 29 June 2002).

Rethinking "Manly" Pursuits: Fishing, Hunting, Climbing

FISHING THE MYSTERIES

BARTON SUTTER

When my cousin Johnny died, I drove west from Duluth through the spruce and tamarack bogs, past the big blue water of Leech Lake, down through the Smokey Hills to Detroit Lakes. Johnny had hanged himself out in Wyoming, but his family had brought his body back to the Minnesota town where he'd been raised. As I drove west against the sun, I went through crying jags, laughing fits, long periods of dreamy reminiscence. This was an awful death, and I was especially fond of Johnny, but whenever I recalled how he looked in the last photo he'd sent me, I grinned. Johnny was standing by a mountain lake in hip boots, holding up a joke fish—a glittering, minuscule trout. He looked amused with the fish, amused with himself, pleased with the shining world at large.

I had taken that snapshot myself, with Johnny's camera, on a visit to Wyoming a few years back. I had stayed for several days in Laramie, and Johnny had driven me around his piece of the West, showing off the mountains of the Snowy Range as if he owned them. We talked nonstop about our childhood, our families, literature and writing, photography and painting, Minnesota and the West. And of course we went fishing. Fishing was a family tradition we'd rediscovered for ourselves in middle age. On a bright, windy afternoon, we tried a dark blue mountain lake, and when Johnny caught that tiny trout—the only fish we got that day—I laughed and snapped his picture. Click. Now he was dead.

It was dark when I finally found the funeral home in Detroit Lakes. I was late, and a small group had already gathered for the visitation—Johnny's parents, Mel and Evie; his sister and brothers; Aunt Margaret and Uncle Ray; the minister; a few friends. We performed a makeshift service there, and, for a reticent family of Swedish-Americans, I thought we did all right. There was no canned music in that room. We simply sang and sang and sang into the silence those Lutheran hymns we'd had by heart since childhood. People spoke openly about Johnny's long struggle with the mental illness that finally killed him. And, one by one, we named the many reasons that we'd loved him. When I took my turn to speak, I said how much I liked his deep, loud voice, his wacky sense of humor, his love for the natural world. And then, my voice breaking, I thanked his parents for taking us out fishing so many times when we were small. That was a weird thing to say, I suppose, in such a somber setting, but my aunts and uncles and cousins understood. For fishing lay very near the heart of who we were as a family.

My dad grew up in Detroit Lakes in a family of six kids. When his parents saw the Depression coming, they sold their house in town and bought a hilltop homestead overlooking the lake, where they could grow a large garden and do some subsistence farming. My dad and his older brother Mel often went fishing together; the lake was handy, fishing was fun, and the meat they brought home helped the family survive. But right from the start, my father said, Mel was the real fisherman. He had the passion. My father was just along for the ride. The truth about my uncle Mel shines forth from an old brown photograph. He looks like a kid in a Norman Rockwell painting, the archetypal Fisherboy—cane pole in hand, straw hat on head, hoisting high his stringer of fish, smiling pride and joy.

Uncle Mel stayed on in Detroit Lakes, became a carpenter like his father, and, whenever he could, fished those lakes he'd known since childhood. My father, on the other hand, moved away and became a minister, a "fisher of men." But most every summer when I was young, we returned to Detroit Lakes, and the climax of every visit came when Mel and Evie organized us all to go out fishing.

We were something of a mob. My dad had begot three kids, and Mel had sired five, so that meant eight brats plus four adults faring forth to test the waters. Looking back, I'm terrified by the logistics. Where on earth did the grownups find twelve fishing poles? How did we ever dig enough worms? Did the women pack a dozen lunches? Somehow the grown-ups managed, and we had so much fun we remembered those outings the rest of our lives.

On windy days we fished off the Long Bridge. When you caught a fish on that wooden trestle, you had to reel it up—spinning, wriggling, flopping—twenty feet through the air. But usually we drove to a resort, rented a pair of boats, and struggled into ugly orange lifejackets, uncomfortable as horse collars. As we launched the boats, Crazy Martha, who ran the resort, called after us, "Don't forget to fasten your crotch straps!" The dads rowed us out, a hundred yards or so, to Mel's favorite spot. We baited our hooks, lowered our lines, and, if memory serves, instantly caught fish. Always. Every time. Big fat sunfish, glittering green and gold.

We were a noisy bunch. The little kids squeaked and squealed as they handled the bait. We shouted "Got one!" again and again as our bobbers disappeared and we yanked back on our rods. That live, electrical pulsing in our hands was a thrilling message from the deep, and when we brought those bright sunfish aboard the boats, we yelled our admiration. Our fathers, naturally, groaned at the prospect of cleaning three dozen panfish. But they admitted that their dirty job was worth it, when, the following morning, our entire tribe sat down to feast on golden fish.

One time, I remember, Uncle Mel took us to another lake, where the panfish were reluctant to bite. But just as we were giving up, in a shallow bay near the resort, my father hooked a five-pound northern. Our excitement ran especially high because Dad was using a rod with a broken tip he'd fixed with glue and fishing line. Would the weak rod hold? Could he land such a heavy fish without a net? As the pike flopped and flailed in the bottom of the boat, I shouted with astonishment. I had never seen a fish that big. I felt as if I'd just watched my father wrestle an alligator and win. I was so happy for him that I hurt. Forty years later, as I recall the bright white heat of that emotion, I feel scalded all over again. The intensity of my boyish love for my father scares me, and I see that, fishing up these memories, I've landed a lunker of another sort, a thought too large to handle. What power parents have! How desperately their children love them!

The truth about my father, though, is that he was a lousy fisherman. He just didn't care. Aside from those trips to Detroit Lakes, he only took us out once or twice a year. And, since he knew very little about the muddy lakes of southern Minnesota, where we lived, we almost always got skunked and came home sweaty and discouraged.

So I took up bird watching instead, a sport and pseudoscience I pursued with passion through my teens. I liked to hike. I liked to camp. And in my twenties I fell headlong in love with the dreamy landscape of canoe country. But I refused to fish. I'm stunned to think how many canoe trips I took without wetting a line. I can even recall several occasions when friends who were catching fish left and right offered me their gear and I declined. Did I think I was genetically defective? Did I think my dad's bad luck was mine? Or was there some sort of Freudian twist to all this? Was I afraid to outfish my old man?

Happily, when I was thirty-five, I suffered a conversion. One evening when my brother was staying overnight with me in Central Minnesota, he talked me into going out to shorefish a nearby pond. To my surprise, we actually caught fish! True, they were only bullheads—slippery, black, and ugly—but they were decent-sized, they fought, and we caught and released one after another, taking turns with my brother's rod, hooting and laughing until the dark came down and mosquitoes rose from the grass in clouds.

It was just an evening of cheap entertainment, but it changed my life, because I couldn't forget that wiggly feeling. I liked hauling up exotic creatures from the underworld. I liked remembering those summer days at Detroit Lakes when I was small, surrounded by my boisterous family. I loved standing out there at the water's edge as the sky went pink and then turned indigo and black as the stars began to blink. Most important, perhaps, that soft evening proved I wasn't snakebit, that I *could* catch fish—if you could call bullheads fish.

So later that year, when my wife and I rented a house on a lake, I bought a cheap rod and reel and began to take the landlord's canoe out on the water almost every day. That's when I got hooked for good. Drifting along in the hush of the morning, mist lifting off the lake like smoke, I felt more naturally myself than I had in a long, long time. I got to know that little lake so well I could paddle straight to the cabbage weed where the fish hung out. Every other day or so, an angry northern would grab my lure and zigzag through the weeds. And more often than not, that meant fish for supper.

Eventually I taught myself to clean my catch efficiently. Northerns are delicious but a little tricky because of all the bones. Before long, though, I was producing neat fillets. And the more fish I cleaned, the more I began to enjoy the process. Or maybe "enjoy" is not the right word. Maybe "honor" or "appreciate" would be more accurate. Because something strange was happening. As I butchered more and more fish, I began to turn Ojibway.

By which I mean: I began to praise and thank the fish. I can't say for sure how this happened; it just seemed the natural thing to do. Those early hours on the lake were filling me with gratitude. I felt a kind of contentment out there that I seldom found anywhere else. I loved observing the gradual changes in the landscape as I drifted through the seasons. I was awake to the world around me—the breeze on my cheek, the odor of mud, sunlight glinting off a turtle's back—in a way I hadn't been for years. And when I managed to land a fish, I was very aware it wasn't because I was a brilliant fisherman; it was just the most marvelous luck. The fish felt like a gift. And so, as I held the pike on the cutting board, as I raised the hatchet to whack it, I found myself saying, quite naturally: "You were a splendid swimmer. Thanks for taking the bait."

When I saw what I was doing—acting out a religious ritual—I was taken aback. But gradually that moment of thanksgiving grew habitual, a practice I continue to this day. I'm not sanctimonious about it. I don't go through any big rigmarole. If the bugs are bad and it's getting dark, I'll just say, "Thanks," stun the fish, and clean my catch as fast as I can, screaming at the skeeters as I go. But I'm well aware, these days, that fishing has become, for me, not just entertainment or a source of food, but a spiritual activity.

Of course it's still a silly thing to do. When I add up the fees for licenses, the cost of lost lures, the miles I drive, it's clear that I'd be better off financially if I just bought a bunch of orange roughy from the store. But I'd sure lose a lot of laughs. When I'm stuck in the gloom of winter, I need only recall the follies of summer in order to raise my spirits. I remember breaking my rod by jerking on a gigantic walleye that turned out to be a tree. I replay the mental movie in which my brother strains to grab his coffee cup off the dock, the canoe tilts, and we sink in the frigid water like Laurel and Hardy. I recall how the neighbor kid snagged a nice northern, watched his cheap reel explode in his hands, then turned to me and declared, "Aw, man. Life sucks." Laughter lifts depression, and scenes like these work better than cartoons.

The humor and stupidity of fishing have gradually become, for me, part of the sport's appeal. After a dozen years of going out on the water thirty days or more per summer, I'd like to think I've acquired a certain amount of expertise. And it's true that I've guided my brother to spots where he's caught fish with his first cast. And it's true that his kids think about me the same way I think of my uncle Mel, as a nice guy who can find fish. But the moment I get the least bit smug, the fish will sink to the bottom and pout. (What are they doing down there, I wonder—eating pretzels and watching TV?) The moment I try to get fancy, I'll cast my daredevil into a tree. Fishing teaches humility.

I've come to enjoy the comedy that fishing entails—even when I'm the clown with the pile of line in my lap. It's the mysteries, though, that keep me coming back. How is it that catching a fish can alter time and space? It takes less than five minutes to land most fish, and yet, like a car accident or a really good kiss, the whole thing happens slowly. Einstein could probably account for this. I can't, any more than I can explain how I've lost the names of so many people I once called friends but seem to remember fish after fish after fish that I've caught, and not just the fish but the setting where I netted them. I can't explain any of this, but, since most of my life goes by in a blur, I'm grateful for these memories—sharp and slow and clear.

Maybe I make too much out of fishing. Maybe it's just a chance to get out of the house. But my kind of fishing—alone, with my wife, with a friend or two, traveling by canoe—takes me through the landscape of my dreams. Far from the grind of urban life, I turn into someone else. I talk to myself, to the hawk overhead, to the wind in my face, to the fish below the surface, beings that I can't see. Winding along a dark river or paddling through reflections, I'm gliding through the landscape, sure, but I also feel the country pass through me. I'm lost and right at home. There's no place I'd rather be.

Fishing has been, for me, one long lesson in ecology, the idea that everything is connected. Last summer, on a solo trip, I caught a golden walleye. I killed it, cleaned it, and left the remains on a rock. Though I hadn't seen one all day long, as I rustled up a fire and started frying fish, a gull appeared in the sky like a star, then descended in circles, crying, to the granite slab where I'd laid out the entrails of the fish. I had seen this happen before, but I was amazed all over again. That bird had come out of nowhere. The walleye was delicious, the white meat fresh and tender. I cleaned up, sat back with my coffee mug, and thought of my cousin Johnny. He'd been fishing with me all summer, even though he was dead. I'd flash on him as a kid in a bright orange life preserver, or hear him snort with laughter, or glimpse him standing by that mountain lake, exhibiting his tiny trout. And again and again I went back to his funeral, where the minister spoke tenderly of Johnny's love for nature, and then the sweetest coincidence, the natural miracle occurred: as we rolled Johnny's casket out of the church to the hearse, V after V of Canada geese passed over us, coming in low, like an air force flyover, calling as they came. That memory made my skin tingle. The evening chill was coming down, and mist was moving in. I pulled on my jacket, launched the canoe, and went out for a lazy paddle. I still had a glow in my belly from supper. As I passed the rock slab where I'd cleaned the walleye, I noticed that nothing was left. Where was the fish? That fish was a bird. I was the fish. The fish was me. And life was a bloody mystery.

Note

This chapter originally appeared as "In Which We Honor Fishing" in *NorthLife* magazine (May 1998), Duluth, Minnesota.

On the Point of a Sharp Hook

JAMES BARILLA

"I've been fishing all my life," my fly-fishing mentor, Myron, once said to me, "except when I was about eighteen. I had other things on my mind for a year or two." He winked at me. I was nine years old, in love with wild trout, fast water, and evening rises. I had no idea what he was talking about. How could anything get in the way of fishing?

A passionate romance and a passionate desire for rivers—different ways of losing the self in another, paradoxes of conquest and dissolution, narratives saturated with illusion. The dream that love never ends, and the fantasy that through fishing we can enter the life of lively water without causing harm, both come true frequently enough to remain poignant, and yet in each there remains the threat of endings, of some revelation in which cruelty and mortality figure prominently.

While it has often been said that blood sports are the proving grounds for male mastery, in fact, the contemporary ethos of catch-and-release fishing is more complex. To catch a fish, and let it go, demonstrates competency, but more important, it represents an attempt to get caught up in lusty, deadly rites while releasing ourselves from the obligation of these sacrifices: the recognition of our own mortality. By releasing our quarry, we sustain the spell of our separation from the whirl of nutrients that once were trout and bears and human beings, even as we wade across rocks that will be there long after we are gone. We seek to retain a sense of immersion, of selflessness, while concealing the fact that the narrative of release, despite our intentions, may end in death.

I fell madly in love with Nicola when I was nineteen, and during long evening strolls, holding hands, during indolent afternoons on the beach, gazing into each other's eyes while reciting significant passages from an anthology of love poems, I found out what Myron meant. Rivers flowed out of my mind. They went dry in their beds, and willows grew tall and shaggy in the arroyos. My fly rod sat in the closet with last year's caddis clipped to a guide, suddenly unloved, like an old pair of favorite jeans kicked off into the corner.

It was a month before I confessed to Nicola that I liked to fish. A lot.

She nodded and smiled, but I felt the watery recesses of my soul had yet to be laid bare, even when I resorted to making casting motions with my arm. I could be talking about any old hobby, a passion for golf or an addiction to video games, when I really wanted her to know what it was like to wade out into a river at dusk and lose track of yourself, how I would disappear for a while in the overwhelming mood of trees dissolving under an opalescent sky.

There came a point in the relationship when the rains came again, when I felt brave enough to introduce one love to another. I suggested a walk along a small stream I'd never seen before, and I mentioned, casually, that in addition to the bottle of inexpensive but impressively labeled Chardonnay, the one wine glass, and the pint of strawberries that I intended to pack, I might bring my rod along too. She said that would be fine, and we set off into the Maine woods on a borrowed moped.

We crossed the brook without even noticing it. A small culvert, with a late summer trickle of amber water glistening between moss-bearded stones, appeared in the midst of dreary slash piles and the somber claustrophobia of a dense stand of evergreens.

I had a vision of this excursion. We would spread a blanket in the sunshine and goldenrod of a meadow, under the mottled arms of an old apple tree. We would sip the chilled wine I had managed to obtain in clandestine underage fashion, and bite strawberries from each other's fingertips. While she lounged and observed, I would amble over to the stream and cast my poetic loops over the murmuring waters. When my line caught the light she would see how graceful it all was, see the artistry of it and be spellbound, enthralled. She would see what fly fishing was all about.

"Look," I said, unable to relinquish the image despite a glance at the uprooted stump that blocked the trail, "I know this isn't great for walking. Maybe we could just go a little ways along the stream. I could show you how I fish."

Nicola swished a deerfly from her face and said okay, if I was sure this was the place. I nodded vigorously. A manic and desperate cheerfulness had overtaken me, marked by the need to whistle. With "We're Off to See the Wizard" burbling from my lips, I pulled out my vest, jointed the rod, and coaxed Nicola, who was wearing sandals, down the blackberry-tangled embankment to some flat rocks. We stooped under the tunnel of trees, our eyes adjusting to the dimness, our skin tingling with the chill. The stream I had imagined was bright and buoyant, almost giddy with light, with pools like fountains of overflowing champagne; this water smelled like fresh-dug roots, the lees of a dark, astringent potion.

"I don't like it in here," Nicola said. I said she was right, it was kind of dreary right here, but there was a patch of sunlight upstream, and maybe even a pool where I could cast. We could make our way up there, and if it didn't open up, we could quit and find somewhere else. She looked at her feet, then at the moss-covered rocks, and nodded an assent.

Mosquitoes discovered and pestered us. We passed nothing worth a cast until we reached the brink of a long, shallow run. In the spring, it might have been a lively stretch of froth and undulating current; now it was nearly still, mirroring the needled boughs above. The rotten crag of an ancient sugar maple stood near some fragments of a stone wall at the head of the pool, its branches long whittled away until only the trunk, festooned with the warts of various fungi, held sway over a small patch of sky.

Fish had never appeared in my beautiful vision of fishing with Nicola, yet now it seemed imperative to justify this excursion with a capture. There would be no golden pasture, no calligraphy of line swirling artfully through the air, but at least there would be a point, a triumph she could observe, however small. I would conjure a wild beast from these waters.

"Stand back," I said.

I flicked a bow-and-arrow cast out to where there was a tremble of current. The bow-and-arrow is not the most elegant maneuver in the fisherman's repertoire; with the rod bent into a curve over my shoulder and the fly pinched between by fingers, I felt like I was aiming a spitball. The caddis dropped into still water. Slowly, as the leader straightened, it found a path between the rocks and swung out from the bank. I glanced at Nicola to see if she was watching its progress. Our eyes met; she was watching me, not the fly. I tried to smile but found the sides of my mouth already taut, clenched in a grimace of nervous expectation. She looked puzzled, almost bewildered, and in that look I knew that whatever vision of grace and beauty I had in mind was not only rendered invisible but negated by the fierce-eyed primate crouched before her.

The fly was riding the folds of the main stream; it was too late to relax, to do more than watch with involuntary fascination as the caddis hovered, spun round and quivered with the current. I felt those movements as little shocks in the already rigid muscles of my arms.

I don't recall the impetus to yank. I do remember the sharp hiss of line and water parting, and the clumsiness of a small and sudden weight, the weight of a fingerling flying through the air. I watched the spell of its sailing, helpless, horrified, as it skittered across some gravel and came to rest with a fleshy snap against a stone. I glanced at Nicola. Slapping at her legs, she looked dazed and harassed, but no more than before. She didn't seem to realize that this wasn't part of the normal routine.

"Was that a fish?"

I nodded.

"It was tiny."

"I know," I said. It was still on my line. I could see it, a black sliver gleaming between contours of dry stone, motionless. I didn't bother trying to keep my feet dry as I sloshed downstream to where it landed. This was clearly an emergency: a fish out of water, making no effort to get back to where it could breathe, while the woman I loved, already dubious, watched the proceedings from nearby.

"Is that supposed to happen?" I dunked my hand and scooped the skinny slip of charcoal and sunset off the rock. The baby brook trout made no attempt to evade my grasp, or struggle its slippery self free of my fingers. It regarded me, unfortunately, with the deadpan gaze and slack jaw of a dead fish. I cradled its body in the water, and petals of blood bloomed and dissolved around the gills, gills that didn't move unless I pushed water back and forth across them. I shook its head. I rolled the body right way up and swam it between two stones.

"What are you doing?"

Praying, imploring, ordering this baby to stay with us, I wanted to say. "He seems a little dazed. I'm reviving him."

I heard some ferns rustle and snap. Nicola was coming over for a better look.

"He's fine," I said, urging some glibness into my voice. "Just hit his head a little." As if in reply, a spasm rippled through the little cadaver, jerking its head.

"There, you see, he's coming around."

"He looks dead to me." She was getting close enough to lean over my shoulder, close enough to see my bloody mistake up close. One more stone and she would get the wrong idea, think I was into killing small things for sport and pleasure. What kind of sick beast would I seem to her then?

I opened my hand to the current. The fish rolled, white belly and scarlet fins streaking the surface, dead eyes just visible under the yawning lower jaw. It drifted away, gaining a bit of momentum, nudging around a stone and slipping into the riffle below. I felt her hand reach my shoulder, steadying herself on the last rock.

"That poor fish was dead, wasn't it?"

I didn't want to face her. I pretended that to keep her balance I needed to stay as I was, watching that belly for signs of life.

"Well, it could revive itself ... " The belly had paused against a branch. It hung there like a candy bar wrapper. Somehow this lack of motion made things final. " ... But it doesn't look good. Yeah, I think it's probably kicked the bucket, unfortunately."

"You're just going to let it float away? You're not going to eat it or something?"

I wished I had never taken Nicola fishing, had never found this place, never insisted on fishing unless the locale was perfect, as I'd envisioned it.

"It's too small to eat."

I turned to look at her. She kept her balance, her hair brushing my shoulder. I could smell the fish slime on my hands, and the coconut of her sunscreen. She looked mortified by what I'd done, by what I'd revealed about myself.

"I mean, I know it's a waste, but something will eat it. It doesn't normally happen like this." She didn't say anything.

"That's why I like fly fishing, because you don't kill the fish. I let them all go."

"It's horrible," she said, and backed away into the ferns.

Sundown, and a ritual of frantic sex and death is reaching its climax in the burnished riffles of the Clark Fork. I can feel the water undulating around my knees, and a small trout is working regularly in my wake. Tiny olive mayflies are skittering in ghostly multitudes across the surface, rising up into the luminous sky and then dropping into the shadows over the water. A spinner fall. By morning, most of them, shorn of mouthparts, empty of fuel, will be dead, reduced to constellations of gossamer wings along the shore. But right now they are dancing with each other, and the trout are involved too, devouring the reckless in a rhythm of small splashes.

I'd like to imagine I'm part of it too, that somehow I've transcended the artifice of nylon and steel, feathers and fur, yet I'm aware that these elaborate displays of courtship differ from anything I would willingly endure. Those jaws lurking just below the surface, promising a quick end—who would choose to run that gauntlet to find and prove their love? Vicariously, we experience this dramatic narrative all the time; it's the stuff of blockbuster films and best-selling novels. But we don't normally pursue romance with the fevered consciousness of the mayfly. Men fall in love with rivers, not fish, because rivers are metaphors of immortality, their essence elusive, their cycles of flood and renewal signaling the possibility of undying passion.

To identify with the quarry is to lose the capacity for sport. The average cutthroat prowling the smooth eddies below Yellowstone Lake is caught over three times a summer, but the same angler rarely stalks the same fish, because a recognizable face in the hand would form the basis for an unwelcome kinship. When I'm fighting a fish, I sometimes think of Old One-Eye, a rainbow trout of unremarkable size and wisdom who inhabited a slow stretch of the Clark Fork River on the outskirts of Missoula, Montana. What made this fish different was its luck and tenacity, for unlike the other twelve-inchers that hung around a series of half-submerged boulders, Old One-Eye had lost half his face. An ugly crater of scar tissue had formed where eye and cheek should have been, the mark, most likely, of an osprey's claw. I imagine this fish dangling from the talon of a sky-bound raptor, the bird's grip just awkward enough that the trout frees itself with a desperate twist, drops through the air and lands with a splash, back in the river but terribly wounded.

I came to know Old One-Eye because I caught him more than once. He and I liked the same stretch of water, and I could recognize the shape of his rise, a disc occasioned by his sideward approach as he lifted his one good eye to appraise a floating object. He was a survivor, but the wound took its toll. While the others were sleek and plump at the end of summer, he was slender, with gaunt pockets between the adipose fins. When the other fish refused my fly, Old One-Eye would grab it. Sometimes I knew it was him and could stop myself from striking. Often, however, I realized too late and would gingerly draw him to my hand, hoping the slack line would release him first.

To imagine Old One-Eye on the end of the line is to wonder whether the steel will prick the rim of cartilage around the mouth and be easy to remove or be embedded deep in the throat, trickling blood. I never had to disengage a deeply embedded hook from his mouth, fortunately, but his disfigured face became familiar, and with that familiarity came a loss of conviction. "Why don't you at least eat what you catch?" Nicola once demanded. I thought about my answer for a long time. I imagined Old One-Eye, the survivor of an aerial attack, reduced to a battered fillet in a sizzling pan, and I realized that I didn't kill him because I didn't need to; if I was hungry and living on the fruits of my hunting skills when he happened along, I'd make sashimi out of him in a flash, with no regrets. But I didn't need to be a predator. I could just as easily simmer tofu as sauté trout.

Aldo Leopold, the founder of the ecological restoration movement, came to a similar point of predatory ambivalence in the midst of a wolf hunt. Faced with the realization that the figure beyond his rifle was a living being, not an object of consumption, Leopold turned to healing damaged ecosystems as an alternative way of participating in nature. I've tried to follow in his footsteps, not just by planting willows along the banks of streams, but more directly. When a drought struck our neighborhood, I was among those who scooped up stranded trout, some tiny, some relatively large, from the nearly desiccated reaches of a small brook. We put them in an ice chest of the sort normally reserved for cooling beer at picnics, and they thumped against the white plastic sides as we sloshed downstream to where the stream emerged again to trickle across its bed.

It was a procedure that felt virtuous, watching our rescued fish slide back into the dark folds of the current and find refuge among the mossy stones. I found myself returning to the place frequently, just to check on them and ensure they were faring well. I wondered if they were getting enough to eat, but I had no desire to confound them with a fly. I hoped that I might replace my predatory forays with an ethic of caring, of live and let live.

Yet my vigils on the bank were often tinged with wistfulness, as if I'd lost something, or fallen out of love. The environmental ethicist Stanley Kane has remarked that what is missing from the experience of restoration is a sense of mystery, of something beyond the confines of the self, beyond the creative interventions of humanity. A trout whose identity is shrouded in the mysteries of a wild river is not the same as one that has traveled there by caring hands, at least in the mind of the angler, and therefore the experience of restoration is not a replacement for fishing. In fact, to restore in this context is to diminish the experience of wildness, of the river as a venue of timelessness and mystery. One has only to visit the Swift River, in Massachusetts, to see what is gained by restoration, and what is lost. Here, countless hours of toil have transformed what was a sluggish and barren meander of chilly water into a virtual arcade of submerged tires, bolstered banks and strategically felled trees. Not coincidentally, the improvement work began after this portion of the river became catch-and-release.[1]

Trout love it; their numbers have expanded dramatically, as have the opportunities for fishing. Here the brown trout rise to inspect the fly from every angle, searching for the telltale curve of metal, as if they have learned to acquiesce to the rules of a game that nurtures their kind for a price. When a fake does fool them, they accept their plight with a minimum of fuss and present themselves to their adversary for a quick release, a gesture that minimizes anguish on both ends of the line.

Because I remember the river as it was, unimproved, mostly devoid of habitat, compromised by a dam and left mostly to its own subtle devices, I do not fish there anymore. Our culture, as the French critic Baudrillard has pointed out, is so saturated with falsifications, imitations, and copies that the inauthentic becomes real, and this is as true of improved rivers as of the local shopping mall. It is the sense of the authentic, that we are wading into something beyond our ken, that generates the illusion of our own immortality, and therefore the signs of human interference register as disappointments, as obstacles to communion.

Men, and perhaps humanity in general, seem to be caught in a quandary of competing desires that have less to do with proving our mastery over other beings than the need for the solace of a paradoxical mysticism.

Ambivalent predators, we find ourselves searching for alternative narratives that will nurture and heal, and lead to faith in a different authenticity than death. It's the hard work of lasting relationships, the faith that keeps the relationship with my wife, Nicola, alive. I try to imagine the shape of these rituals as I wade into a stream and prepare to cast my fly.

Note

1. One might argue that since this stream exists below a dam, its characteristics are artificial from the start, and indeed, the Swift's character is shaped by the inability to recruit stones from upstream, by the introduction of non-native fish, by the temperatures that result from releases from the bottom of the dam. Yet I believe this is a question of intentionality, and therefore a question less of the degree of human impact than the nature of human designs. Consider the restoration work on Michigan's Pere Marquette River, where sand traps have been installed to improve habitat damaged by poor forestry practices, and the principle still applies.

I Love the Single Deer Path

TIMOTHY YOUNG

I love the single deer path
winding into the wet, tangly night woods
where nocturnal squirrels,
and whip-poor-wills
usually fly tree-to-tree.

The Hunter walks from the known
to the not-known,
and water drops from leaf to leaf.
The humus grows moist,
still and womanly.

Hunters walk in life, and take a life in order to feed life. Or they hunt ritually to express how meaningful the hunt has been for the continuation of life. To me, now, it is that simple.

I grew up in a household of conscientious and disciplined hunters. My dad and uncles taught us to respectfully hunt rabbits, ducks, pheasants, and ruffed grouse. We ate what we hunted or gave the game to old-timers who relished the meat. However, when I reached my mid-twenties, I quit hunting. Ten years later when my son was born, I returned to hunting. This story tells of some of the reasons why I left hunting, and then returned to it.

From my earliest years I lived with the imagery and a sensory awareness of meat as food. My father was a meat cutter in St. Paul. One of my uncles raised cattle on his farm where I spent many weeks each year with my cousins, and where Dad taught me to hunt ducks on a back field pothole. My grandfather had been a South Dakota cowboy before he married and moved to St. Paul, and he told me story after story about being a ranch hand on the prairie. His brother drove cattle trucks from South Dakota to the South St. Paul stockyards, and whenever he stayed overnight with us in "The Cities," I slept on our family couch. When he had returned to Dakota, my bed glowed with the rich odors of diesel fuel, animal hair, manure, and cattle feed—soulful, intimate, and unashamed odors.

Because my father worked in a city supermarket, some of my earliest memories are of meat behind a glass counter, a sawdust floor, sparkling knives, and whirring band saws. The year I began high school, Dad began to cut up deer for hunters. This extra work brought in money for our large Catholic family, and we developed a small, seasonal, family business. The profits paid for Catholic school tuitions, family necessities, and most important today, hunting land in northern Minnesota.

When we processed the deer, my task was to skin the hide and then, before sending the quarters of meat to my father, to burn off with a propane torch the sticking, hollow winter hairs. Small tulips of smoke would lift into the darkness above the garage's loft joists. I will always have memories of the odors of burning deer hair, cold venison, a doe's suet, and a stag's dirty rectum. In ten years, I skinned over seven hundred deer. I know well how a deer's bared ribs feel, and venison hams and slippery bone knuckles.

As the deer skinner, I had to listen to each man tell his kill story—the area where he hunted, the type of cover, the quiet spot where he stood, or sat, or leaned against a tree, the way the deer approached, its head movements and antler swings. Having heard ten stories like it the day before, I was unimpressed. Yet, because they were customers I never stopped a story. My restraint was not due to respect. I was an arrogant, introverted college boy among working-class men. Because these men craved to tell their stories, as men should, and because most were never taught how to tell stories without a beer in hand, we were at odds.

One evening as I worked, an especially obnoxious braggart insisted on prattling his expertise to me. I realized that I could probably chase him from the garage if I turned the radio from familiar, if disliked, "hippie" music to the more mysterious classical music. I found a classical music broadcast from St. John's in Collegeville, Minnesota. When I returned to my work he continued to babble. However, his words became fewer and fewer, and his babble more disjointed. Finally there was silence. I looked at him. He stared at me like a deer in the middle of a highway. Then I became conscious of the sensory impact of my work. I moved and worked intently with a slender knife in my hand. I wore a bloodied, white meat-cutter's smock. A half-skinned deer carcass hung from darkness, and its gaping chest cavity exposed inner ribs and blackened blood clots. Yellowish fat wrapped the backstrap, and folded skin flaps held soft white belly hair while a full, pink doe teat dripped milk to the concrete floor. Through the garage's odors swirled the sublime music of *Jesu, Joy of Man's Desiring*. The hunter froze in the midst of this beauty and blood. We, two, stood at the edge of something larger than we understood. I could feel, though just barely, the spiritual power pulled into the garage by the music. He left the garage, and I went back to work.

I did not understand why such moments were so spiritually powerful. But I felt something. From the first day I hunted with a shotgun and refused to kill living pigeons just to practice my skills, I had questions about my own hunting as "death-dealing." What happens to animals when they die? When I kill one, am I evil? Is there a spiritual realm beyond this earthly world where the souls of animals and humans continue in a different existence? I instinctively knew that the hunting and killing of another creature was a core activity for the human soul, and for human sustenance. A hunt is not a hunt unless there is the threat that something dies. The conscientious hunter must accept this. He cannot turn away and absolve himself from his participation in death-dealing; neither could I dismiss as irrelevant the vitality I felt while hunting. I sensed that there was a spiritual dimension in hunting. My father and uncles accepted hunting as normal, worldly behavior and let it be at that. Our parish priests could not answer my question about hunting's value to the soul. At this time I knew nothing about the attitudes of indigenous hunters, and I did not yet realize that my questions were part of a human being's maturation process. I felt only guilt and confusion.

A blanket of bravado wrapped most of the hunters I knew in a state of psychological denial. The commercialization of the natural outdoors, the merchandising of the hunt, the "Welcome Hunters" signs on taverns, the governmental need to manage prey kills with bag limits and seasons, the citification of the American culture and the rising awareness of ecological disasters stifled respectful discussions and encouraged blustery and defensive rhetoric among the hunters I knew.

My exposure to relevant writings and opinions was limited to *Time, Sports Afield,* and *Outdoor Life*. I never felt rooted in a literary or religious system that could give authority to my beliefs. The only serious hunting literature I knew of glorified killing or tried to emulate Hemingway's work, which I did not yet understand. Nor was I interested in the work of someone who would kill himself. I secretly loved poetry. Yet I could not find serious poems about hunting or its relevance to spirituality.

Recently I looked back at some of the poems I wrote when I was twenty-two years old. My themes were the same themes I work with today: that beauty and the natural world depend upon a great recycling of living things; that the shadow side of rational, scientific thinking is fear, especially a fear of life's unpredictability, which then creates in us a craving for safety and purity. I was clearly seeking a spirituality in nature and an awareness of my own natural soul. The invigorating excitement of a hunt added another layer of meaning to my quest. The word "soul" darts into and through many of those poems. They speak of "absolutions," "rebirths," and "*holy* wisdom." Here are a few lines pulled from a rambling, poorly written poem:

> In the city my soul becomes an urban skeleton,
> an abandoned infant decaying in a can.
>
> In the pine I am a fresh intruder, but still wet with stink.
> Animals flee because my mother has not licked me clean.
>
> Like a devil, tension is cast out
> by the holy spirits of the elements
>
> and the weight of sunlight through
> needles and boughs forces age into
>
> the crevices inside me, fills those cracks,
> and packs oldness into my inner spaces.

I knew almost nothing of the spirituality of other peoples. After twelve years of Catholic education (many as an altar boy), after two years at the University of Minnesota where my Catholic education was challenged, I began to question my religious background. In my professed faith and spirituality, I was completely Catholic. However, the natural world was infusing me with a deeper sense of spirituality. I was struggling to understand.

I had been taught that animals did not have souls, yet I felt differently. When my friends and I killed animals, we would go drink beer "to celebrate." Such drinking numbed my soul. I wanted to be "one of the guys," and I desired a natural participation in the wilderness. These are contrary desires, and I was not ready to choose one over the other. I wanted wisdom and I wanted to avoid the burden of my dilemma, which was, "Why does one soul have to kill another?"

The human body senses things deeply while on the hunt. My physical senses were heightened, and as an undisciplined person I had difficulties coping with such heightened awareness. My eyes saw more textures, movements, subtleties, and patterns. Scents were innumerable, and poignant. The range of sound was vaster than anywhere else. More importantly, the possibility of enormous silence gave my life a larger potential. My soul quaked under the burden of its growing awareness. It was exhilarating and expansive. To survive and not yield to numbness or fear, my awareness of myself had spread out. But it also condensed. The pulse of awareness, inward, outward, inward, outward, was as important to my vitality (maybe even more so) as the pulse of my physical heart.

TIMOTHY YOUNG

I also learned that nature itself kills with brutality. Cruelty, ugliness, and violence are as inherent as beauty, harmony, and peace. For instance, the most courageous struggle of a small vole may be while it dies in the elegant goshawk's talons, as the beak rips the vole's flesh so fledglings can eat.

To comprehend the beauty and vitality of nature, one must hold the paradoxes of life and death in the soul. During the depression, my dad's family revered pheasants, squirrels, the wild woods because they survived by legal hunting and poaching. After the depression, they continued to hunt and fish. They valued the unique beauty of the birds, the cleverness by which animals survive, and the challenges of an intensely lived life. The ritual gathering of food and the ritual killing for meat reminded them of the precariousness of life. They enjoyed being alive. My brothers, cousins, and I inherited this enjoyment, and I attempted to bring its vitality and paradoxes into the hidden life of my poems.

In the autumn of 1972, Dan took a break from his wildlife management studies. I, too, had decided to skip a quarter or two of college. We spent six weeks hunting and fishing every day, along the lake, down the small rivers, and through the flooded swamps where Dan also set traps for muskrats. Mostly, however, I scribbled lines about my inner struggles. Here's a stanza from one poem.

> I place my faith upon his skill
> to supply the meat for our meal.
> As helmsman, I guide
> the canoe, not by stars, but a thrill
> of the hunt and forced survival
> in the wild. Only death may decide.

I knew intuitively that to mature we had to understand the meaning of survival. Of course, we would not starve, but we had to act as if we might. My poetic exaggeration was a form of a prayer. We were unconsciously ritualizing the sustenance of life through the hunt, and the hunt ritualized our family's story, which in turn is the story of humankind's struggle. But as a young man of the times, I was also asking, "Am I serving death or life?"

In 1972, death and dying were imagistically ubiquitous. I needed to differentiate between the death-dealing in hunting and the death-dealing in warfare. The Vietnam War had brought more violence and threats of violence to my life. Violent images spread through the media. My friends returned from the war with rage and wounds. In 1970, my own conscription was prevented at the last minute by the severe acne that had erupted on my body. I was sent home from the induction center only minutes before the other draftees began their trip to boot camp. Despite the "summer of love" enthusiasms of the rock-'n'-roll sixties, death energy had overtaken the iconography of my generation. Janis Joplin, Jimi Hendrix, Mama Cass, and others had died as drug-using rock idols. Altamount had followed Woodstock. Because I secretly dreamed of being a poet, John Berryman's notorious suicide while I sat in class across the campus gave death a greater imprint on my mind.

A more personal tragedy occurred in 1969. My nine-year-old brother, John, died from a brain tumor. We witnessed his bouts of pain on the family sofa, and we watched him suffer through his surgery, radiation treatments, and final miseries. In 1972, my family still lived in the shadow of the simple question—Why was an innocent child taken by such a cancer?

My poetic exaggeration (only death may decide) was, therefore, not an exaggeration to my soul. My immature abilities focused upon the power of death. Young men do, naturally and instinctively, move toward the edge of chaos where creativity emerges and death challenges their survival. Whether consciously or unconsciously, young people need to prove to themselves and to their community that they are capable of sustaining their own lives and that they are useful to others, as protectors and providers in the ever-changing motions of life. To do so they have always gone to the edges of their culture or community.

Dan and I were in the darkness of emerging manhood. We had entered the great, tangly night woods of northern Minnesota, and each of us followed a path into the unknown. We were looking for the elemental experiences that would enliven us with hope, healing solace, explanations, and reverence. Years later, the Mayan spiritual teacher Martin Prechtel explained that we were trying to initiate ourselves into the mysteries of life. Our instinctive attempts were noteworthy, but futile. Self-initiation is not possible. An older, experienced person must consciously lead a young person to the edge of life's mysteries if the maturation is to be life-enhancing and life-respecting. When faced with the enormous power of death, the young one will emotionally shut down or begin to move toward insanity if not accompanied by an elder's attention. Here's a small example.

Four years before Dan and I went north, I took one of my neighborhood buddies on his first hunt for grouse. I coached him how to shoot a shotgun, taught him the hunting protocols, and told him about the pleasures of upland bird shooting.

We never did jump a grouse that warm September afternoon. On the walk back to the car, we kicked up a cluster of frogs from the grass beside a marsh. My friend, frustrated that he had not killed a bird, emptied his shotgun on one jumping frog after another, while cackling with an insane look on his face. His neck muscles shivered in his adrenaline rush. His body crouched like a frog. His eyes bulged crazily. This mild-mannered eighteen year old gave in to an emotional greed for violence. My friend had never touched the carcasses of food animals. He did not know the coldness of the animals' death, nor did he honor the life-giving aspects of meat. I could not teach him that which I learned much later—a grateful relationship with nature must be sought and acknowledged when one hunts another life. If no relationship exists, then the activity is merely extermination. Extermination is the act of fear and greed gone amok. I saw this vividly. It disgusted me. I vowed to never again take a gun-toting neophyte into the woods.

After that autumn alone with Dan, I slipped into the lifestyle and habits of the very fellows I had scoffed at earlier. I became numbed and oblivious to any spiritual aspects of hunting. In August 1974, I went with the "neighborhood guys" to shoot clay pigeons and prepare for the upcoming bird season. We were drinking beer, bragging to one another, insulting each other, and just being boorish. We were not paying attention to the natural environment. All we wanted to do was avoid the police and game wardens and enjoy our guns and beer. Such mindlessness is a prelude to danger and destruction. As we walked toward the backwaters of the Mississippi River, a wood duck suddenly lifted from a pothole. One friend, always a heavy drinker, swung toward the duck. His shotgun arced toward my head, and I dropped to the ground just before the blast. Had I not dropped, he would have killed me. I became so angry and rattled that I abandoned my friends to the slough and drove my car home. Four of them had to cram into a small pickup. I had lost any meaningfulness for hunting. I had almost died. For a week I replayed this near-death experience, and I remembered every violent, uncomfortable experience I had while hunting. I had no sympathetic confidante, and I knew I no longer wanted to hunt.

Over the next ten years, I finished my undergraduate studies, returned to graduate school, wrote poetry, and performed it with musicians. Even though I did not hunt, I tried to feed the hunter inside me. I tracked beauty. I hunted spiritual texts, art, music, and quiet moments in the natural world. Often I went alone to libraries, concerts, and museums. I needed the time to seriously examine beauty without obligations to a date or companion. These experiences felt similar to those I had found in the woods, in a canoe, or walking in the snow.

Ironically, the path I followed with poetry folks led me back to rather than away from hunting. I became generally familiar with Native American beliefs. I listened to politically active Native Americans, and I paid close attention to the hunting lessons woven throughout their indigenous stories and literature. Meridel LeSueur, the octogenarian grandmother of the peace and justice community, also taught that life is circular and cyclical, and she encouraged the publication of my poetry, which included the imagery of hunting. From the work of great American poets I collected poems about the depth of the hunting experience, and I was surprised to find that when they did write about hunting, almost all of our great poets wrote reverently.

When I followed poet Robert Bly into the woods for one of his men's conferences, I found new teachers, white men, who had learned traditional knowledge from indigenous teachers. Among the various teachings about the soul, those men spoke of valuable lessons to be learned from proper and respectful hunting. Some teachers taught in the manner of indigenous people, through stories rather than lectures, through poetry and music rather than training videos, and with conscientious rituals that revered and honored all living things.

Then I married. My wife brought her young son to my house, and we had another son. A deep force swelled into me. I felt it as a father and a hunter. My five brothers, my father, and other relatives continued to hunt, and they continued to invite me to deer camp. They knew that I felt the need to return to hunting. Something primal aroused me. I now had a deeper context for hunting rituals.

When I needed a respite from fatherhood, I began weekly trips to parks just outside the Twin Cities. I practiced stalking deer, just as the tracker John Stokes had taught at a conference. I tracked animals in the snow until I found their winter homes. With the help of guide books and gathering bags, I more urgently identified various plants and animal life. In all kinds of weather I sat for long periods and sensed the incredible richness of life, and the teeth of death that fed the living.

Then I bought a bow. I wanted to know the quiet hunt a bow hunter knows. I wanted to learn the patient skills, and to be physically and spiritually connected to the woods and my prey. Martin Prechtel taught that a young man must fall in love with "the goddess, Nature" so that he can throw his soul's great expectations to her, rather than onto his human lover. Large expectations and fantasies become debilitating to the human women men love. The intensity of one's inner life can never be fully handled by another human, despite our romanticized expectations of marriage and relationships. Nature, on the other hand, can accept and handle all the emotional turmoil and grief a man is willing to express. The grief and attention we offer to divine Nature is like candy to her. When we give our grief and tears we feed her, and cleanse ourselves. I was falling in love, again, this time with the goddess. I understood the importance for myself, my sons, and my wife.

After a couple years as an unsuccessful bow hunter, I wondered what I was doing wrong. I knew only a few conventional bow hunters, with whom I did not feel comfortable. So I continued to read, for practical and spiritual advice. One book contained the songs Pueblo peoples sang for success on a hunt. One song addressed the cougar as the king of hunters. No melody was indicated in the book. Since I lived near the Como Park Zoo, I visited the cougars one Monday morning in October after I had practiced at the archery range. The morning was bright, and the high sky was deep blue. On their den that had been built from concrete and shaped as a mountain crag, the cougars lounged in the sun. The zoo was quiet and nearly empty of people. The cougars watched me intently. I spoke to them quietly about why I had come to visit. Then I sang a melody that popped into my head, and I adapted words from the Pueblo verses.

> He comes alive, alive, he comes alive.
> He comes alive, alive, he comes alive.
> He comes alive, alive, he comes alive.
> The lion of the north, he comes alive.

That evening I left for northern Minnesota. As I approached my freeway exit near midnight, I saw the shadowy form of a medium-sized cougar seated beside the freeway. As I turned onto the exit, the cat ran into the woods. I felt stunned and awed. The next day I returned to verify what I thought I had seen. Sure enough, I found a cougar's spoor and some scat on a sandy logging road. I took it to be a sign, and I hunted that area. After only a half hour, a doe came running directly at me. I drew my bow and waited, but at about thirty yards, she saw me just as I was about to shoot. She spun behind a pussy willow bush. Directly behind her a very small yearling came running. Behind it, something else moved in the tall grass. The yearling spun with its mother, and they ran across an open swath of clear-cut forest. I was not capable of shooting a deer fleeing at full speed. But it didn't matter. It was a timeless moment. A surge of awe and grace and joy came from a deep place inside me. I had been a part of a cooperative hunt with a cougar. Just as the indigenous teachings had promised, I had received assistance from the "king of the hunt." I had ritualized my prayers in song. The cougar had chased two deer toward me. One for me and one for him. However, I was still too inept to fulfill my part of the bargain. As my gratitude swelled, I gave loud, grateful apologies to the unseen predator. A spiritual dimension opened to me. I felt blessed by unseen forces. Three years later, Martin Prechtel told me how to ritually make amends to the spirit of the cougar through prayers and offerings. When I prayed directly to the cougar spirit, when I gave it ritual gifts, I found another layer of meaning to the hunt—reciprocation. In the spiritual dimension reached through ritual, the intention of one's heart can penetrate ignorance and pain, and can heal the soul.

That winter, I finally bought a Winchester rifle and an over-and-under Winchester double-barreled shotgun from a retired military friend. The following November, I joined my brothers and family for the annual deer hunt. My brothers knew of my years studying the literary and spiritual dimensions of hunting. They knew that I went with groups of men to study the outdoors, old folktales, and the cultural traditions that bonded men in the unique matters of the soul. They knew the "cougar story," and they knew that I now valued the hunt as a sacred, ritual activity. I had told them personal stories about my experiences with the teachers, the poets, and the men at these conferences. They were aware, yet wary, of my attitudes.

In the week before Opening Day, my brother Jake told me he was glad that I would join them. "You're the only one who'll understand this," he said. "Whenever I've killed a deer, before I do anything else, I put a little pine bough in its mouth, and ask it to forgive me, and I thank it for becoming food for my family." His words were almost the exact words I had read in hunting stories from tribal peoples. He knew instinctively what I had needed to learn. I had also learned that the act of putting grass in the deer's mouth was an ancient, medieval German rite. Jake was pleased, but not surprised, when I told him.

As I rode north with a second brother, Tony, we jabbered about our families, politics, union work, and current events. After a conversational pause, he said: "Well, umm, I'm really glad you're hunting with us. You're the only one who'll understand this, but whenever I kill a deer, before I gut it, I whisper a little prayer of thanks to God . . . and to the deer. I know the church wouldn't see it as appropriate, but I think it's necessary." I sat quietly, grinning, and said, "Thanks, for having me with you." Two of my five brothers, who had been hunting together for years, were praying for the souls of their deer kills, and they were embarrassed to do so openly, although neither was ashamed to pray at church.

On opening morning, I hunted from a deer stand on Dan's woodland. He has hunted more than the rest of us, and his knowledge about the animals, their behaviors and habitats, probably exceeds the cumulative knowledge of his brothers. He has also had a ritual that he performs after every successful hunt. Once he kills a large animal, he cuts out its tenderloin, "the love muscle," from behind the heart cavity. As soon as possible he fries it and shares small pieces with whoever is near. His "tradition" is similar to the rituals of tribal people, who ritually cook and eat a piece of the heart to honor the individual animal and to absorb some of its personal strength.

After a few icy hours sitting in the tree stand, I heard him shoot. I waited awhile, then crawled down and walked through the snow toward him. As I came from behind a stand of fir trees, I saw him kneeling over a fallen deer, whispering in its ear and patting it on the head. I allowed him time to finish, then noisily approached as he began to gut the deer.

I did not say much to my brothers about what I knew of their private, sacred rituals. But over the next years, I purposely spoke about my own need to pray for the animals, for thanks, for the rituals and the family bonds that are strengthened by our hunt. Always, they nodded and said little, but I could see that they agreed.

Finally, on one hunting trip after I had been hunting with them for a few years, I prayed openly over a deer, the way one teacher had taught. It felt awkward to me, but my brothers gave me respectful space for my words of gratitude.

That evening, after a first course of venison tenderloin cooked in butter, we ate a dinner of buttercup squash and green beans from Dad's garden, a pork roast from Jake's farm-raised pig, and cake sent north by Mom. Uncle Bill came by to visit, and while he and Dad talked in the cabin, the rest of us, including my other brothers, Joe and Chris, went out beneath the spectacularly lit Milky Way and smoked cigars in the dark. We spoke together of the spiritual dimensions of the hunt. They even let me recite a few hunting poems. Brothers, a brother-in-law, nephews, all of us speaking in reverence for the mystery of life's basic law—Something dies, so another can live.

Fathers and Sons, Trails and Mountains

O. ALAN WELTZIEN

I remember my excitement that sunny June morning in 1994 when Bill Neighbor Jr., Galen Stark, and I studied the big 3-D model of Mount Baker in the old North Cascades National Park Headquarters building in Sedro Woolley, Washington, excitement about finally stepping onto Baker that didn't even dissipate at the foggy, dripping, crowded trailhead. Once above timberline I didn't feel crowded off the moraine, and by early evening we made camp at 6,500 feet. As it was only a couple of weeks before Midsummer Night's Eve, we'd several hours to cook, eat, laugh, reminisce, and pan the view to the east, south, and west. We carbo-loaded and donned extra layers after the sun sank, though the rounded rock outcropping held its heat a while longer. I stared at Baker's peak, and the foreshortened view, its dimples and folds, filled my vision, the yellowing light pulling the dome, more than 4,000 feet higher, almost within arm's reach. Looking up, I anticipated the morning; looking southwest, I surveyed my past.

On at least two occasions in recent years, I have climbed old friends of mountains with a pair of old friends. Washington's North Cascades National Park boasts as fine and uncrowded a climbing region as any in the lower forty-eight states, and over most of these serrated jumbles and jutting blue points loom Mount Baker and Mount Shuksan, as famous a pair of mountains as any in the Evergreen State. Mount Shuksan, at 9,127 feet, is one of the most photographed nonvolcanic mountains; its west profile, whose angular rock ridges offset steep snowfields and glaciers, appears regularly on calendars. Mount Baker, at 10,778 feet a perennially white queen, northernmost of Washington's five volcanoes, and portal of northern Puget Sound, dominates the mountains' west edge, invitation to the riches beyond. Baker's dome gleams like a vanilla sundae, curves to fit our picture of volcano. While I knew Shuksan secondhand from calendars and coffee table books and a laminated placemat pen-and-ink sketch acquired during childhood, I grew up summers staring at Baker's southwest profile, and at my folks' beach place I'd frequently flick my glance across bay and foothills to make sure it hadn't moved. Long ago I knew its detailed shape with my eyes closed, so when in the summer of 1994 two of my best friends proposed climbing it, I jumped at the chance.

I only hiked with my father a few times, and he never considered climbing, yet his earlier example probably set my feet in motion. Dad was a member of Boy Scout Troop 125 in North Seattle for several years, and in more than one summer in the mid-1930s, he spent two weeks at Camp Parsons, along the shores of lower Hood Canal—Puget Sound's longest finger. During the second week, Troop 125 would backpack up in the rain shadow quadrant of Olympic National Park, and decades later he regaled me with stories of "Poopout Drag," a steep piece of trail well-known among Puget Sound Scouts of that period. He discovered he liked hiking. During adolescent summers he and one buddy, Harry Christopher, day-hiked frequently in the South Fork of the Snoqualmie River drainage—where U.S. 10 and, later, I-90 changed the landscape. Dad wore hobnailed boots and carried a thermos of hot cocoa for the end of the hike, and I imagine he and Harry often saw no one, though when I hiked these same trails as a teenager with more than one friend, we found other parties along the lakeshores or on the mountaintops. And now hundreds of boots may pound any one of these trails on a summer day.

In later life, Dad took to rivers with two buddies, and though they got splashed aplenty and Dad was tipped out at least once, they never worked too hard negotiating rapids, leaving that to the young bronzed guides at the steering oars. His river hat, which hangs in my garage, bears the embroidered names, in sundry bright colors, of his western rivers and summers. They hurtled rapids and poked in side canyons without their spouses, and though women were present on these trips, the absence of wives defined them as much as the canyons. They wore grizzled face hair and fewer clothes, and I expect their conversation grew both raunchier and more philosophical than it would at home. They tasted various pleasures greedily, and in those earthseams where rivers tumble, their close friendships tightened further. Dad always considered one of them, Frank Milam, as close a male friend as he'd known since his days with Harry Christopher. "Big Frank," our family ophthalmologist and an incredibly bright man, ate, drank, and read too much, buying new hardcover copies of whatever books in whatever disciplines whenever he wanted. Years later, since Dad was a Lewis and Clark aficionado, I took him and our older son on a trip down the "Wild and Scenic Upper Missouri River," which proved our last river trip. Three generations of Weltzien men retraced Lewis and Clark campsites while drifting 149 miles through Montana's White Cliffs and Badlands.

If my father's scouting and later river experiences gave me, early in my life, examples of a life lived in intense moments outdoors, my friends on this volcano climb embodied the value of expert local knowledge. On Mount Baker we pitched Bill's Northern Face tent on a snow flat with its flap opening just south of east, and sitting up in our bags, from the aperture we surveyed the fierce young North Cascades from the British Columbian border south to Glacier Peak. Bill named the peaks in the Picketts he'd climbed, and he and Galen pooled their knowledge of at least a couple dozen other names and profiles. A duet, they chanted a litany of names. As chief wilderness ranger of North Cascades National Park, Galen had learned "his" park's topography in some detail; he has hiked most of the Cascade Crest Trail between the Canadian border and Stevens Pass, though he has climbed less than Bill. Now and then I recognized a peak, remembered a climb; sometimes stories trailed after names. I sat between them, privileged by their familiarity, privileged in our lifelong friendship.

Bill, Galen, and I have been friends since kindergarten, if not before. Bill grew up four blocks south of my house and Galen, four blocks north. Bill and I lengthened together in Boy Scouts and later attended the same college. Earlier, Galen and I weathered Indian Guides with our long-suffering fathers, and we used to chant together the pledge, "pals forever," as we romped in one another's houses or on overnight camps. They were my two primary day-hiking buddies during high school and college summers, my Harry Christophers, who adored the Cascades at least as much as I and who didn't mind sweat-wet shirts and handkerchiefs. Bill is on the faculty of the University of Washington Medical Center; Galen recently retired as a career National Park Service employee who worked mostly in Mount Rainier and North Cascades National Parks. Both of them lived elsewhere but returned to Puget Sound and resumed their hiking and climbing in the North Cascades. My ninth-grade English teacher told us we'd be lucky if we found and kept two or four friends in our lives, and though at the time his stringent definition puzzled me, I've come to understand it. Bill and Galen and I roamed together through childhood, adolescence, and adulthood, and we climb a mountain together to remind ourselves of our braided journeys and bedrock intimacy.

As we sat watching the sunset's shadows hurtling across the snowfields, thinking about summiting the next morning, the tent felt like a male space. Our wives rarely backpack, and no wives were invited on this trip. Does the span of our friendship alone explain our preference? Or is it the rare opportunity to be together, temporarily casting off our identities as husbands and fathers, once again driving then hiking high in the Cascades, eager for an airplane view? These trips up neighboring peaks bring out, in certain moments, the boys inside the men, a familiar tale, but I think it's more a matter of us each taking the measure of our lives from the others. We serve as compass bearings for the others just as the two mountains provide geographical compass bearings; studying one another, we remind ourselves of the paths we've taken, and I remember trips when we took turns leading up a glacier. Our voices change less than our hair color and hairlines and faces, and we take stock of the difference. The grain of their voices recalls my past and lights the lengthening, sunlit path between then and now. No women, not even wives, play more than bit roles in this business.

By dawn the cloud cover had risen to about 20,000 feet, and as the morning wore on, it thinned out and sun shone before we summited. Galen broke trail, punching steps and occasionally pausing to blow, and we skirted crevasses and approached the swirling lip of Sherman Crater. Close up, Mount Baker belies its chaste, distant serenity, even its symmetry, because its crater vents south of the summit dome and approximately 900 feet below it: a violent, mysterious chasm where thick tendrils of steam part momentarily to reveal glaring ice, and bare bands of pumice offset the deep pile snow carpet. We pushed up the dome and met three climbers and an unleashed black Labrador beginning their descent, the dog providing the contemporary comment as we trudged up the final hillock, called Grant Peak. Though by the 1990s, standard routes up Mounts Baker and Shuksan resembled mass promenades that mask the notion of personal pilgrimage—they had become a glacial Pilgrim's Way across England's South Downs to Canterbury—we had, miraculously, the summit to ourselves for a quarter hour. We three grew up atop Clyde Hill, which rises more than 300 feet above Lake Washington, just east of Seattle, and now we'd managed a bigger hill, the highest we'd climbed together.

We studied the details of lower Shuksan, just to the east, which we would climb seven years later. Bill and Galen again gestured farther east and southeast across the jumble, recapitulating their roles as cicerones. Standing above our native sea, surveying our past, we picked out some Puget Sound islands and shoreline to the west and southwest. We had feasted rather than fasted, and I experienced no vision quest, yet my vision was richly autobiographical since we scanned decades as well as miles.

Later, Galen and I eagerly descended, post-holing in softening snow, sometimes pulling the rope taut against Bill, encouraging him to make bigger steps. At camp we struck the tent and weighed down our packs again, and a few hours later, we sat in the Baker Street Grill, in Concrete, Washington, toasting with microbrews before the food's arrival. As I remember the trip, those sunset hours with Bill and Galen stand out probably more than our windy minutes on the summit or rest stops at the crater. On Mount Baker we three inhabited a space all our own, one defined by our diverse mountain stories and altitude and body odor. Altitude and heavy packs helped us simplify and clarify. Inside that bubble of time, breathing thinner air, we halved our age and shed our family relationships, retaining only that oldest one that glues us. We raised our glasses that evening because we had bagged the peak, yet that fact did not lessen the spirit of communion, of being enduring friends on this mountain together. I had finally come close up to my summer mountain, touched it with ice ax and crampons, stood at the other end of the binocular views of childhood. My father followed trails and shunned elevations above timberline; Bill's dad fishes and Galen's golfs. Our generation, at least annually, prefers more demanding settings.

Though I like more demanding outdoor adventures than my father, I, like him, learned to love the woods and hiking trails in the Boy Scouts.

Dad used to talk about how much his Scout years meant to him, and he hoped it might stir similar yearnings in me. I had joined Troop 600 with several friends in Bellevue; this suburban Scout troop numbered more than thirty boys and five patrols, and parents and the First Presbyterian Church supported it generously for years. I camped quite a few weekends, also attended Camp Parsons for several summers, rose through the ranks, and attained Eagle Scout and a couple of Palms. More to the point, I hiked and backpacked and canoed, and since my family did not, that has made all the difference. The Scout year climaxed with the fifty-miler trip, and two summers I canoed Lake Coeur d'Alene and Priest Lakes in the Idaho panhandle. Another summer we backpacked tracing a horseshoe, east to north to west, around Glacier Peak, and though the cloud cover stayed low and it rained and I carried Dad's gruesome old army pack, I learned to stay dry some of the time. I never pooped out on endless switchbacks because I grew up stubborn and because I half-sensed my enchantment with the North Cascades, a magical alpine region in the lower forty-eight states. For some reason, Dad's teenage fondness for day-hiking has burned more deeply and steadily in me, for I have grown up obsessed with mountains, first the Cascades and Olympics, and later, many other ranges.

The following summer I carried a better pack, and we hiked fifty miles south along the Cascade Crest Trail, beginning at historic Naches Pass, northeast of Mount Rainier, and ending at White Pass. As if to compensate for the preceding summer's rain, this July the sun glowed as we passed the length of Rainier's east profile. An older Scout, I walked last in line, usually staying in the CCT rut, picking up the occasional trash and stopping to drink views whenever I wished. Excepting occasional swarms of mosquitoes in lake basins, we hiked in nearly ideal conditions, and that sunlight on arms and legs wedded me to this range. By this time I'd begun day-hiking in the Cascades and had fallen deeply in love with walking in the mountains, despite my chubbiness and lack of fitness. During the summers of high school and college, I relied on a small group of buddies similarly beckoned upward. The fact that we didn't invite any girls comments mostly on our lack of social development, though it too has made all the difference.

Once I reached adulthood I'd slimmed down and, by age thirty, begun running. I'm no marathoner but run regularly each week year-round, in part because my fitness permits me those mountain trips in summer. Over the years I've learned to temper my haste and pause to identify heartleaf arnica and mountain penstemon and phlox and myriad other perennials I can't seem to remember the names of. My wife, who loves gardening more than most things, has helped me look just beyond my boots. So I study foreground as well as distance, but my eagerness to gain altitude has not diminished; if anything, the increase in years has sharpened it. So many mountains to hike or climb, most of which I'll never reach. Thoreau would be disgusted with me as I know so little how to saunter; instead, I am driven and don't like to pause for long rests or blows but only to slow the heart rate and note the range of colors or understory or grain of rock or views across or back down.

I can never study mountain profiles enough, never step onto them enough. I know by now that I'm little more than a mediocre climber, but several rock and ice climbing classes have decreased my fear of exposure and increased my confidence on standard routes. Those classes numbered as many women as men. I'll never be any big wall guy, for instance, because I lack the skills and passion and fearlessness, but I have climbed far more mountains than Dad ever wanted to, and I always set my sights on other summits. I like to summit—I like that verb—and crave that occasional chance to pretend the omniscient point of view. I know that for myself and many, a spiritual yearning prompts my sweat. Aspiration feeds inspiration. Ascending a peak, I often feel as William Wordsworth atop Mount Snowden or John Muir atop Mount Ritter in his beloved Range of Light, but sometimes I react as Henry Thoreau did atop Mount Katahdin, which only underscores the gift of awe mixed with humility. For myself at least, I scramble or reach handholds as a pilgrim, and peaks beckon as high altars.

As I age, however, I increasingly wonder whether my peak-bagging ego represents much more than a belated, deflected form of machismo. I've never strutted my stuff on a court or field because I never had anything to strut, but I can hike and scramble fast and long, and I don't climb on routes that will paralyze me. Perhaps two forces—in psychological conflict—drive me.

My family and I have lived in Southwest Montana for a decade, and my feet have come to know many of its local mountain ranges, each bearing its own name. I like cross-countrying below and above timberline in Southwest Montana with altimeter watch accurately calibrated and 7.5 series USCGS quad maps in hand. So much space and so few people, many of whom hunt in autumn—it's a secular religion—and tie flies in winter, awaiting spring and summer caddis and mayfly hatches. Probably a majority of my students, women as well as men, hunt antelope, whitetail or mule deer, or elk, or fly fish, yet most of the camps and stories belong to men. My quarry is less palpable. Foot traffic in the ranges in our corner resembles what I imagine it to have been in Washington's Cascades in the 1950s or earlier. I can repeatedly reach above 10,000 feet in more than one range within our county, Big Sky's largest, where cattle vastly outnumber people and where, up high, I can see in all directions fifty or more miles on a summer day.

With one colleague and friend—a very bright theoretical physicist who publishes prolifically and who climbed with big names a generation and more ago—I retreat for two to three days every summer to the Beartooths, the rooftop range of Montana, usually for one route and one peak. It's become a tradition in recent years, one we anticipate acutely, and though I enjoy his wife's company greatly, as he does mine, neither of us could imagine them joining us, apart from the fact that they lack the fitness and stamina to do so. Craig has always climbed without Karen just as I climb without Lynn. I, at least, imagine my loss, the new forms of intimacy Lynn and I miss exploring, but I also know the gain when Craig and I push onward unencumbered except by our quiet proximity. Craig is quiet and restrained, I am garrulous and too often unrestrained; yet with him in the Beartooths I quiet, trying to match his taciturnity, the generous space between utterances, and listen more to the Beartooths beyond my labored breathing. We treasure our intimacy, only interrupted, if at all, by a few mountain goats or a couple of high-altitude anglers or fellow climbers.

Maybe in adulthood I've been compensating for my childhood and teenage lack of athleticism through retreats to the mountains, and if in these occasional retreats I have "simplified, simplified," I've turned to male company to do so, just as my father faced downriver with two male friends. Surely this intimacy between Craig and me would change form completely were our spouses with us. Among our friends, I can think of only a couple of women who enjoy pushing beyond lakes for views, but again, this may comment more about our circle of friends than anything else. In ten years I cannot remember when women have stood with me gazing beyond the verge, the farthest range.

My own family, unfortunately, replays gender clichés I initially sensed growing up. In her Girl Scouts years, my mother never backpacked, since her family couldn't afford to send her to summer camp, and her troop never hiked otherwise. When I was young, she enjoyed car camping with the family though she rarely hiked; she wasn't drawn to the woods, though one memorable summer day, years before they married, she allowed Dad to persuade her to hike up Mount Pilchuck though she wore saddle shoes and though he never leveled with her about the mileage or elevation gain. While Girl Scouts backpack frequently nowadays, their primary public image remains cookies in increasingly diverse shapes and fancy packages. My stepdaughter never joined Girl Scouts, and neither she nor my wife (or mother) likes to get far "out there," particularly if it means steady sweating.

My wife likes to go on short hikes if we'll see lupine or Indian paintbrush or arrowleaf bitterroot or wild irises, and if the trail is not steep. My stepdaughter went "camping" with other girls in high school, but that meant parking a car between sagebrush and timber and building a fire and chatting half the night or more, and probably included boys and alcohol occasionally. She has never shown any interest in being in the woods or on mountains, and she says she hates hiking. Our older son, by contrast, has day-hiked with me, and though he is now almost my height, he poops out on long steep stretches, and I let him. Mountains don't fire his lungs and legs. Our younger son has hiked shorter trails, but like my wife, he dislikes the prospect of many miles. I can remember a rare family hike a few summers ago, though it led, after only a mile and a half, to a warm mountain lake with a sandy beach.

In recent decades, climbing has become a mass sport, and women sport chalk bags and place protection as naturally as men—there is no reason to believe woods and mountains belong any more to men than women because they don't, yet that has been my experience. Do alpine settings encourage behavior stereotypically or particularly male? I remember my bemused chagrin when I read Erik Erikson's *Childhood and Society* long ago and absorbed his gender report: boys build towers and skyscrapers with wooden blocks, while girls build circular enclosures, like medieval walled cities. Despite this archetype, I've always suspected that my peak bagging involved something more than male conquest of feminized landforms articulated so thoroughly by Annette Kolodny, among others; something more like religious retreat, in which I've been fortunate in my choice of company of one or a few male friends. As Don Scheese, a fellow mountain fanatic, remarks in the beginning of his recent book, *Mountains of*

Memory, "I have often wondered about humanity's love of prospects, the desire to see the world from the highest point around. Is our love of mountaintops a pre-Romantic phenomenon, an atavistic practical instinct?" (11). Do I fulfill my alpine desire not only out of Romantic ardor or religious disposition but out of some deeper biological instinct as well?

My wife likes to grow plants, her eye reveling in the scale of seedlings and flower gardens; I sometimes forget to look closely near my feet in my eagerness for a 360-degree view. An unabashed descendent of the Burkean gospel of sublimity, I scramble and climb in the religious spirit of a John Muir because I seek prospects near and far that reduce me to insignificance. I want the earth spreading below me as though I'm a momentary subject of special revelation. Such desire for epiphany belongs to people, not men, but in my story men seek those big broad views, that accomplishment and ecstasy, more insistently than do women. My wife, probably a more religious individual than I, seeks self-fulfillment on a less ambitious, less arduous scale. The difference is a matter of degree.

The desire to climb is not inherently male, of course, because the twentieth century has, in many cultures, created myriad opportunities for women to become alpinists far beyond my calibre, and women have followed up on those opportunities. A rural citizen, I neither hang out in climbing communities nor frequent climbing walls, both of which women do; I don't lead climb, as some women do. I once taught Arlene Blum's *Annapurna: A Woman's Place* (1980), and I relish their double entendre T-shirt mantra, "A Woman's Place Is on Top." I like to share peaks, but with very few others and despite the fact that as I age, the gap widens between experience atop mountains and language; my words fall further behind my feelings. I hike or climb, a combination of conqueror and anchorite, and ponder the mixture. I don't think it's gender specific, yet in my own little story, men seek the mixture more regularly or more predictably than do women. The mixture negotiates between self-aggrandizement and self-effacement, pride and humility. If men historically have sought out the mountains, or other natural settings, for purposes of conquest, an element of pilgrimage has accompanied them as well. Borrowing John O'Grady's title, they have walked more insistently than women, it seems to me, as pilgrims to the wild. Sometimes the pilgrimage has been primary, as prophets and writers have testified through millennia, and I like to believe I've always sought something far greater and more elusive than the commonplace rush of victory. Above all, being in the mountains invites introspection that, in my case at least, leads me to question what it means to be male as well as merely human.

In the summer of 2001, Bill, Galen, and I finally got organized to climb Mount Shuksan via its gently sloping Sulphide Glacier, a broad, relatively easy approach. For this climb, Galen was on duty along with his subordinate, a climbing ranger named Eric, and a seasonal novice, John, new to these mountains: they wore their National Park Service summer uniforms. They hiked to base camp on the Glacier a day before Bill and I, joined by Matt Bennett, stepson of a college buddy, arrived. We left the usual crowd of vehicles and climbers strung out along the Forest Service road end, and along the bushhogged trail in the midday heat Matt, Bill, and I got wet quickly, sweat drops falling off eyebrows and chins. Blackflies had divebombed us while we were packing up, and once we were on the trail, they flew at us in swarms whenever we paused. Eventually we topped out on a ridge, climbed a short pass at timberline, leaving the blackflies behind, and began chunking across snowfields lining the Glacier's lower reaches. I led, my pack weighing less than Bill's, and in the lengthening afternoon, I spotted Galen, who was looking out for us—our personal welcoming committee. Old friend watching for old friends. We pitched Bill's same roomy North Face tent near the Park trio's and cooked on "their" rock outcropping. One party of three, two men and a woman from New York, camped below us. Before Bill, Matt, and I cooked, we walked, with Galen, west and a bit higher near another party's campsite to drink the panoramic westward view at sunset. A couple of guys had claimed this prime spot. Here above 6,000 feet, Mount Baker filled our field of vision, sprawling left and right in rough symmetry; theatrical yellow sunlight, a widening corona, streamed round the volcano and pulled us forward, our voices subdued. Three middle-aged guys and one undergraduate: four male figures, tiny and clustered, in the corner of a sprawling Albert Bierstadt canvas.

The next morning the summit pyramid grew big and close as we skirted its base, stowed our lighter packs, and organized our ropes for climbing the southeast ridge. We had reached the base fairly quickly; thereafter, since Eric had decided to join the two ropes into one supporting six climbers, two with little or no experience, we slowed way down, a deformed beetle with six pairs of legs, only two of which (including Eric leading) moved at a time. We crawled up the ridge, spending most of our time waiting for someone else to move, and my impatience mounted. As I awaited my turn just below a steep block of rock, that trio of climbers from New York "simul climbed" past us, reaching the summit and beginning to descend the gully when a transient heat lightning episode stopped our progress, within a hundred feet of the summit, and turned us down the gully. I wish I had summited with them rather than sitting too much along the ridge, waiting to climb. Summits matter to my ego, and it's tough to fall just short, though our views from the upper ridge stretched almost as widely. On the ridge I had noted their tight, stretch guide pants and rock shoes—they dressed more fashionably than we—and envied their grace and speed. Their system of commands worked smoothly as one would belay, for ten or fifteen seconds, the next. I also noted that the woman took the middle position on the rope; both ascending the ridge and downclimbing the gully, she neither lead climbed nor cleaned the route.

In my small bundle of mountain stories, men—myself and a few male friends—occupy all the ground; at best, women in other climbing parties flicker across my peripheral vision and disappear from consciousness. In the absence of wives, for example, we shed certain constraints, along with deodorant and razors, and concentrate on our desires rather than negotiate them with theirs. Maybe we curse more or tell dirty jokes or remain silent longer. On rock or snow, those desires reduce to the task at hand, which for me becomes an occasion for introspection, as in a religious retreat. When I gain one of those all-directional views and my muscles relax and breathing slows, I happily shrink into a tiny point, a secular monk for a short while; Bill or Galen or Craig stand, at the most, in my peripheral vision. It remains a quiet communion, one stripped of most of the conventions of social intercourse, as my eyes slowly pan.

Works Cited

Blum, Arlene. *Annapurna: A Woman's Place.* San Francisco: Sierra Club Books, 1980.

Erikson, Erik. *Childhood and Society.* 2d ed. New York: Norton, 1963.

Kolodny, Annette. *The Land Before Her.* Chapel Hill: University of North Carolina Press, 1984.

———. *The Lay of the Land.* Chapel Hill: University of North Carolina Press, 1975.

Muir, John. *The Mountains of California.* 1894. New York: Penguin, 1985.

O'Grady, John P. *Pilgrims to the Wild.* Salt Lake City: University of Utah Press, 1993.

Scheese, Donald. *Mountains of Memory.* Iowa City: University of Iowa Press, 2001.

Thoreau, Henry David. *The Maine Woods.* New York: Norton, 1979.

Wordsworth, William. *The Prelude.* New York: Norton, 1979.

mothers, fathers, sons, wives

AS BIG AS THE WORLD
Imagination, Kindness, and Our Little Boys

JULIA MARTIN

> *Mommy, why do people have guns?*
>
> *Are there really bombs in the world?*
>
> *At our school there's a game called British Bulldog, and when you play, you say: "Army, army, in the grass, how many bullets up your arse!"*
>
> *Mommy, when I'm big I'm going to make a big safe place for all the animals and the people to come in so that baddies can't shoot them. I'm going to make that safe place as big as the world.*

It's parents' evening at the playgroup. The gathering of mothers and one father sit in a circle on little chairs in the beautifully painted room. The teacher is telling us about gnomes, fairies, water nixies, and undines. She says these beautiful metaphors help the children to understand that the whole world is alive and peopled, that they are participants in a sacred environment. When the formal part of the evening is over, we drink tea and start talking about our little boys: what to do about their fierce aggression and their love of guns, police, jail, chopping, and chain saws. The language of gnomes and fairies, it seems, is not enough.

For boys growing up in contemporary South Africa, the conflicts, bullyings, and polarities that they are likely to encounter in the playground reflect structures of violence and dominance in the broader community. Although the long years of colonialism and apartheid may be over, the impact of their tyranny endures in the inequalities and prejudices that mark our bodies and minds. Violence against women, child abuse, racial fears and racial violence, the seemingly unbridgeable divide between rich and poor, the frightening escalation of violent crime—our new democracy takes shape in a fractured and polarized society in which many people are armed against one another, and most of the violent crime is committed by men. For many, many little boys in our country, the role models for masculinity tend to be aggressive, coercive, competitive, domineering, and unemotional. There is, in a sense, no zone of safety. Even in the alternative preschool that our privileged, middle-class children attend, these values find expression in play.

But perhaps there's more to it than this. Like many other feminist women I know who have given birth to sons, I find that the experience of raising a boy has made me reconsider what I thought I knew about gender and socialization. I have found it bewildering, perplexing, and exasperating to see our wonderfully loving, inventive, boisterous, tool-using, humorous, earnest, musical, beautiful, beloved son, Sky, now just turned five, acting out the fear-aggression gestures of macho patriarchy. His twin sister, Sophie, though fierce and sometimes furious in her own way, doesn't do this. She's "not really insterred in guns and stuff." Could it possibly be something inherent, an innate "nature" that makes boys be boys? For anti-essentialists, the idea is a difficult one. Yet when one of the mothers from the playgroup told me that around age four the testosterone level in a boy's body doubles, I was pleased to have found a way of interpreting these aspects of Sky's behavior. There's a certain relief to be had in thinking that it's all biology.

But of course it can't be. Although some child-raising writers have produced convincing and helpful guidelines for parents that assume a testosterone-based theory of boys' development, there are many others who question the primacy of this emphasis, focusing rather on the influences of interpersonal and social experiences.[1] Instead of engaging with these debates, I would like to assume that parents and other caregivers are in fact powerful agents in the lives of our sons, and to think further about how we are equipping them to inhabit the strange world of the twenty-first century. Much has been written in recent years about the fundamental role of fathers and older men in shaping a boy's experience of masculinity. In South Africa, colleagues in psychology and criminology have researched these aspects specifically in relation to the climate of violence in which we are living.[2] This emphasis on the importance of male role models is clearly valid and valuable, and I am sure that the quality of fathering is, increasingly as a boy grows older, a crucial, irreplaceable and defining factor. But this is not my main concern here. As a mother, I want to write from my experience of mothering. As a feminist, I would like to continue the conversations about raising boys in which feminist mothers in other parts of the world are engaged.[3] In particular, I will be writing about what I call little boys, preschool boys (three to six year olds), since this is where my own attention is at present, and because I understand that the first seven years or so of a child's life are crucially formative.

For the generation of small boys growing up at the beginning of the twenty-first century, the global crisis of environment and development will set the defining conditions in which they learn to be men. In the South, people are already experiencing the erosion of their environments, and within the lifetimes of our children, unless current trends are reversed, all beings will share the devastation in some measure. In South Africa, the medium-term impact of industrialized Northern lifestyles will be felt very keenly. Predictions for global climate change indicate that in fifty years the greater proportion of our country is likely to be desertified, with much of the territory rendered uninhabitable for human beings, and thousands of plants and animals becoming extinct.[4] Entire ecosystems will be erased. The abundant paradise of wildflowers up the West Coast that our family visits every spring, the numberless fields of daisies, nemesias, and gazanias that the children have known since they were born will all be gone. The flowers are an indicator for the rest of the system. When they are gone, the rest of us are also endangered. As this suggests, one of the clearest lessons of the present crisis has been the unmistakable insight that nature can no longer be imagined as a separate realm of plants and animals. Our human labor, our culture, our words, our lives are all inextricably interwoven in the big earth system of non-human beings and processes.

Yet the polarizing ideology of dualism persists. Living in South Africa, the hierarchic oppositions that are the fruits of our history are particularly evident, painful, and difficult to ignore. In spite of the critical stance that I try to take in relation to the discourses of hegemonic power, it is clear that as a tertiary-educated, middle-class, "white" woman, I enjoy privilege and opportunity that is far beyond what is available to most of my countrywomen. My husband, our children, and most of the people we know well are similarly positioned. The experience of living in this situation is particular and localized, yet the relations of power, disempowerment, and inequality in South Africa reflect in microcosm the global dichotomies. North and South, so-called "developed" and so-called "developing" nations—these terms point to a division between rich and poor in which about 20 percent of the world's people consume 70 to 80 percent of the earth's resources. As the South African experience illustrates, this divide is one that runs through all of our societies, although in different proportions, separating a class of rich, globally networked consumers from members of the poor social majority, who are outside the global circuits.[5] While a minority consumer class may, in the short term, enjoy the benefits of this inequality, its impact is devastating for the earth and for most people. As our ecosystems rapidly degrade, violent conflict is everyday news, global corporate interests operate across national boundaries according to laws unsanctioned by democratic process, and poor people are increasingly being deprived of the resources for their survival. In this context, young boys (perhaps especially in South Africa) inherit an ideology of masculinity that reflects in microcosm the global relations of power that are poisoning the earth. Aggressive force is promoted as the privileged form of agency; living ecosystems are objectified as a storehouse of natural resources for human progress; and the world is polarized into North and South. Their socialization into the dominant models of manhood prepares our boys to be competitive and aggressive, equipping them fairly precisely to be part of the problem of the ecological and social degradation of nature rather than agents of healing and transformation.

How can we intervene in this socialization, we who share to some degree the riches of the global North? In this time of unprecedented ecological and social crisis, how can we help them grow up, instead, to be men who are compassionate, imaginative, wise, and brave? After thinking, talking, and feeling my way around these issues for some months, I am even less certain than when I began. In our family, we talk about questions like this often, I've read what I can find on the subject, asked ten friends who are mothers of little boys to write about their sons, and tried to listen closely to Sky and Sophie and their friends and to be more aware of my responses to them.[6] At present, two things seem rather obviously clear in all this. First, there exists no formula or book to tell us how to get it right. There is no switch for kids. Since the twins were babies, we have consoled ourselves through tantrums, fears, and sleepless nights with the reminder that "we're just going to have to muddle through." Although some parenting authorities may insist on the absolute truth of one or another model, living with growing, changing children offers a reminder that our best theories are always provisional and conditioned. At the same time, the daily needs of love and food and play and boundaries ask us to keep making decisions, wiping noses, and acting on the basis of what seems sound in the circumstances. Walking this groundless path is easier and more delightful if we're not too serious about it, or too attached to the idea of results.

Second, I suspect that if little boys (not all, but many) want to bash things, find sticks, launch missiles, point things, charge around, make a noise, we can help them find safe, spacious, and socially congenial environments in which to do this, but there's not much we can do to stop them. Nor would it be helpful to try. These energies are wonderful and powerful and they need to find varied expression.[7] It's perhaps what *else* they do, their capacity for activities and attitudes that are not gender-specific, and the understandings that they bring from these to the experiences of being a boy that seem to be crucial. So for boys who are unlikely to grow up to be either hunters or warriors in the old sense, we need to be able to nurture intelligences and sensitivities that will equip them for living with kindness and imagination in the nondual nature of living beings that is their home. Kindness, compassion, and imagination are not inherently gendered, although in patriarchal cultures they tend to be associated with "feminine" rather than "masculine" orientations. If our boys are to find happy alternatives to the more destructive models of male identity, they will need to rediscover their vast capacity for expressing these key faculties that are their birthright as human beings. So in what follows, after considering some of the structures by which these are often denied, inhibited, and subordinated, I would like to suggest some possibilities for remembering and nurturing our small boys' imagination and kindness, and the intelligences of heart and mind that may accompany them.

I'll begin with the near-pervasive language and visual imagery that are used to structure our boys' view of masculinity in South Africa through globalized marketing and the mass media. In the mainstream stores, clothes for children from about the age of three reflect gender divisions in very stereotypic terms. Girls may wear many colors, but especially pink, with an emphasis on the frilly, the floral, the twinkly, and numerous images of Barbie. For boys, colors are darker and generally less radiant (green, gray, navy blue, and camouflage); there are clothes that mimic sporting and fighting gear, Superman pajamas, and battleship socks. In many toy shops the polarized options are even more of a caricature. Girls can dress dolls, put on make-up, and play house. For boys it's vehicles and guns: convincing-looking machine guns, handguns, grenades, and the absurdly pumped-up muscle-bound testosterone-laden action men that stand in for role play. Television confirms the divide, while repeatedly presenting images of men in aggressive, often violent situations. Reinforcing the effects of an already violent and polarized society, the values that are presented to little boys in this way through marketing and the media are problematic in two main respects.

First, and most obvious, the content promotes a narrowly stereotypical set of options for masculinity, with a strong emphasis on competitive, macho, and often violent conflict. Little boys are thereby encouraged to see their relation to the world in dualistic, even antagonistic terms (boys versus girls, us versus the enemy, me versus the others, and so on) which habitually situate the male subject in positions of power and superiority over others.[8] One effect, among many, is to deny agency and subjecthood to all "others," preparing the boy for being a man who is short on empathy and good at dissociation.

The discourses of militarism offer an extreme, but instructive example of what is involved. Whether the war is waged against people or nature, animals, or ecosystems, similar habits of mind seem to be repeated. I recently read for the first time my grandfather's regimental history of the Durban Light Infantry, which clarified some of these features of the military man as expressed through the male line of my own family. As colonel of the regiment in wartime and school headmaster during peace, A. C. Martin tended to embody the family's figure of patriarchal authority (albeit in its generally benevolent aspect). His history of World War I is full of a young man's stories of adventure, dangerous exploits, and endurance. But by World War II, photographs of Grandad look hard and unflinching: short mustache, chest and shoulders braced, and the unsmiling gaze of the military alpha male. The stories are harder too. But still, the joy and glory of it all, the "fine" and "gallant" soldiers, the "indomitable men," the "great generals," the heroic language used in the lead-up to the Battle of Alamein, and the grand excitement of the earth, which shook for six hours without ceasing (285). Hadn't anyone read Wilfred Owen or Sassoon? But what stays with me is those bayonets: the image of young men under machine gun fire, running forward, who "carried on with the bayonet, whilst they too, sang the Zulu 'Ye-ye-ye-li mama, jioleli mama,' a lusty, gutsy song popularized by the Umvoti Mountain Rifles during the campaign" (291). I remember the chant from my childhood, recited out of context by my father, my father whose mind was blown open to poetry and psychosis by Alamein, my father who read literature and played the piano, whose imagination wouldn't stay quiet. There must have been many such people. In *The Eye in the Door*, from Pat Barker's trilogy about World War I, one of the characters somehow always misses in bayonet practice. They have to rush at the sack and stab it and slash it, and he can't bring himself to do it because he can't not imagine what he's doing. He can't achieve the act of dissociation that is required of him as a soldier. And yet, in a study on boys, M. Miedzin describes "decent, nice guys" who have been involved in war and seem not "to have thought of the people whose lives they took as anything but abstractions" (297). Following Hannah Arendt, she connects this lack of empathy with lack of imagination and argues that what she calls "the masculine mystique," in combination with the detachment and denial made possible by modern technology now make it "even easier for decent men . . . to commit horrendous acts of violence" (297).

If dissociation or failure of imagination are among the required qualities for a successful military man, they also seem to be part of the mental equipment required for functioning unquestioningly as a late modern civilian. In *The Lives of Animals*, J. M. Coetzee's character Elizabeth Costello describes Nazism in terms of the failure of imagination of its perpetrators. Yet her purpose in doing so is to extend the argument by seeing an analogy between this particular (and definitively evil) failure of imagination and a collective denial that is practiced daily throughout the civilized world. Her subject is the pervasive and unremarked abuse of nonhuman animals, and the act of dissociation that fails to engage with the continuing horror of drug-testing labs, factory farms, and abattoirs. Then in the case of Jews and others, now in the case of animals:

> [T]he killers refused to think themselves into the place of their victims, as did everyone else. . . . In other words, they closed their hearts. The heart is the seat of a faculty, *sympathy*, that allows us to share at times the being of another. Sympathy has everything to do with the subject and little to do with the object. . . . There are people who have the capacity to imagine themselves as someone else, there are people who have no such capacity . . . and there are people who have the capacity but choose not to exercise it. (34–35)

In view of the concerns of this essay, we may extend Costello's contentious analogy from animal abuse to the ecological crisis. Here the degradation of human and nonhuman others that so-called development requires can be seen as a similar failure of heart and of the capacity for imaginative identification. When little boys are socialized into the dominant models of masculinity, we should not be surprised if they grow up to deny the suffering of nature.

If the hegemonic masculine stereotypes teach little boys othering and dissociation and thereby the shutting down of imaginative identification, a further aspect of this material is possibly more insidious. Quite simply, mainstream marketing and the mass media position the young child as a consumer rather than as a maker of things and meanings, and privilege centralized "truths" over local, regional, situated knowledges. While appearing to confer power on the little boy, the effect of this is, ironically, disempowering in many respects. He learns to disregard the local, the particular, the handmade, and loses touch with the specificities of his own place in order to reach for objects and ideas that are globalized and mass-produced. Caught in the bright promises of commodity gratification, he learns to forget his own capacity for embodied manual skill, and once again for imagination.

Television and the toy industry are particularly effective in dulling the mind in this way. Among many writers who have examined their negative effects, Joseph Chilton Pearce uses brain science to argue that the major damage done to children has not primarily to do with content. Rather, its damage is neurological. Among other effects, "television floods the infant-child brain with images at the very time his or her brain is supposed to learn to make images from within," providing both stimulus and response (165). Chilton Pearce goes on to argue that this is far more serious than not being able to daydream:

> Failing to develop imagery means having no imagination.... It means children who can't "see" what the mathematical symbol or the semantic words mean; nor the chemical formulae; nor the concept of civilization as we know it.... They can sense only what is immediately bombarding their physical system and are restless and ill-at-ease without such bombardment.... Having no inner imaging capacity leaves most of the brain unemployed, and a child who can't imagine not only can't learn, but has no hope in general. He or she can't "imagine" an inner scenario to replace an outer one, so feels victimized by the environment. (167–68)

Similarly, the proliferation of toys means that children are inundated with objects that "don't stand for something, but already are" (169). Again, this means that the crucial metaphoric-symbolic learning is stunted, and the imagination impaired.

At this point in writing I have just heard that the husband of Nowethu, our child-minder, and father of five children, has been shot dead. A few months ago it was her nephew, and before that her sister. "How are the children?" I asked the neighbor who called. "Just sitting on the beds, crying," she said. In the communities where many parents are unemployed, or if they're lucky, working-class, little boys may not have toys or new clothes, but they do see men with real guns shooting people dead. In the face of this pain, my concerns for our middle-class sons seem trivial. And yet their stories are inextricably connected. Oh, rage and weep for them all!

Here are now some simple ideas for developing our privileged and embattled little boys' skills and awarenesses in other directions. Although the most dangerous and destructive features of global culture seem to find their acutest expression in the socialization of boys into men, girls are obviously affected in different (if related) ways, and the suggestions I make apply to girls and boys equally. In fact, the interventions I would like to make aim particularly to recognize and foster capacities and qualities that are not gender-specific. These may sound more prescriptive than I intend. My hope, primarily, is to give our children a taste of something "better," more satisfying, than the wares of globalizing corporate development. In this, it is clearly not enough for us to tell our little boys how to behave, or what to do. Rather (and sometimes with alarming precision), they learn from watching and doing the things we do. The work and play of our bodies, minds, and hearts are the inheritance that instructs them. These, and the empowering knowledge that we see them and trust them; that their openness, sensitivity, and love are recognized and valued.

First, and perhaps fundamentally, it is important to nurture our little boys' imagination. If they are to grow up to be men who can see beyond the promises of development, to perceive the nonduality of nature, and work to heal the ecosocial crisis, we need to empower them to be creative, intelligent people who can believe in their own capacity for agency. At the level of imaginative play, this could begin with removing television from the home, telling and reading lots of stories every day, learning poems and singing songs, gradually reducing the quantity of shop-bought toys a boy has to play with and showing him how to make stories, plays, and books out of minimal things. I've found that whenever I speak to mothers about small boys and violence, the topic of guns almost always arises. Although anything, even a piece of toast, can become a gun, we don't want replica toy guns in our house, and when anyone gets very enthusiastic about gun-play we remind them that guns kill people—like Nowethu's husband, her sister, and her nephew. When it goes beyond merely reproducing the ready-made elements of consumer culture, imaginative play can become embodied in manual skill. Sharing with children the infinite possibilities of sand, clay, stones, water, paper, wood, wire, cloth, string, plants, wool, food, scissors, crayons, and other found objects is a wonderful gift in which the body/mind learns the powerful joy and confidence of being a maker of things.⁹

Related to the idea of imagination as symbolic or metaphoric play is the equally crucial capacity for imaginative identification with human and nonhuman others that we may call sympathy, empathy, or even compassion. If the stereotypic ban on boys showing strong feelings other than aggression ("men don't cry," and so on) brings with it a subordination of this capacity (that girls, by contrast, are often encouraged to develop), this seems to reflect a characteristic emphasis in Western culture. Anne Klein describes this as follows:

> To the extent that personal creativity and individuality are more valued than relationship in the West, to the extent that autonomy is characterized as the pinnacle of psychological and ethical development, there is the implicit suggestion that caring and a relational style of identity make one less than one might be. Thus, compassion is framed in opposition to more singular forms of selfhood. (Klein 101)

Associated with this is the devaluation of motherhood, and the requirement that boys, as they grow into men, distance themselves from the kindness and relatedness that feminine identity tends to imply.

And yet, pervasive though this view of autonomous subjectivity may be in Western culture (and in the globalized media that derive from it), well-established alternatives do exist in certain non-Western models. Klein shows that Buddhism, specifically as expressed in Tibetan culture, offers important resources for developing an unoppositional view of self in which compassionate relationship "expands rather than limits the potential of its bestower" (110).[10] Similarly, in his explication of the Mahayana concept of *bodhicitta*, His Holiness the Dalai Lama discusses how practitioners might develop this state of mind/heart in which "kindness is combined with the highest intelligence" (Gyatso 16). Central to this is a practice called the exchange of self and others, which involves "putting ourselves in the place of others, and putting others in our place" (104). Living in a culture that sanctions coercive power and related forms of violence, we need to find ways of applying such practices. The heart's capacity for imaginative identification is surely one of the most valuable things we can share with our children, particularly for those of us living in the global North. For example, writing in the lead-up to the 2002 Earth Summit, the authors of *The Jo'burg Memo: Fairness in a Fragile World* insist on the necessity of "wealth alleviation" if we are to reduce the "ecological footprint" of the rich and alleviate poverty.[11] How can we, or our children, begin to meet this challenge unless we change the way we think and feel about "others"? Together and daily, we need to draw on models that prioritize both imaginative and practical forms of play and work that value kindness, sharing, giving and listening, and affirming our relatedness through sympathetic imagination and the heart.

In Mahayana Buddhism, the compassionate heart is inseparable from insight into interdependence. In this context, the basic premise of ecology, that "everything is connected to everything else," extends to a non-dual view in which all selves or entities arise in mutual dependence and are nothing in themselves. For young children growing up in the shadow of ideologies that promote competition, autonomy, and dualistic thinking, some version of this insight can be very empowering, even progressively subversive.[12] One source for this could be in literary or educational materials for children that implicitly or explicitly seek to evoke an environmental consciousness. In South Africa, aside from the majority of works in this field that promote rather moralistic conservationist ethics, a few texts do (with varied success) attempt to raise awareness of interconnectedness.[13] Internationally, the most inspiring work in this area for young children that I have read is called *each breath a smile*. This picture book by Sister Susan, based on the teachings of Zen master Thich Nhat Hanh, constitutes a sort of guided visualization in which the young child is led, through attention to breathing, to the recognition that "all life is breathing with us" (n.p.). Instead of promoting an intellectual knowledge *about* interdependence, such works aim to evoke an experiential, embodied knowledge *of* it through regular practice.

As this last example suggests, in enabling our boys to be heartful and intelligent men, it seems important to develop a lively communication between intellectual understanding and ways of knowing that may be called nonrational or intuitive: embodied and situated knowledges. This is particularly clear with regard to an education in interdependence, where place may be the most significant teacher. Various writers have noted the crucial formative influence of childhood familiarity with wilderness for many future environmentalists.[14] And in his ecological critiques of mainstream American education, educational theorist David Orr argues strongly for a reorientation toward a focus on place.[15] Whether or not our little boys are lucky enough to live close to wilderness, they need to perceive and learn to love their embeddedness in a specific location, and their interrelatedness with the plants and animals that share it. How do you feel when the seasons change? Where does our water come from? What are the native plants? Do you know which ones are edible, and which poisonous? Can you see birds from where you live? Following these and other questions could be the beginning of a bioregional education, wherever one is living. With our own children we have found enormous pleasure in gardening: planting seeds, seeing vegetables grow, tending, wondering, eating, sharing, composting. It helps if you have some earth to dig in, but even a window sill is space enough in which to discover our continuity and connectedness with other growing things. Even in the most urban environments "wildness won't go away" (Snyder 15). Instead of perpetuating the dualistic habit by taking a stance in opposition to the impoverished models of masculinity promoted by the global media, learning about place offers a reorientation that may be more profound. Instead of seeing himself primarily as an autonomous self at odds with others, the boy may come to experience the broader locatedness of his "self" in the growing, changing, living nondual nature that grounds all culture: the neighborhood, the bioregion, the earth, the wild. In remembering his continuity with this community of beings, he may come to recognize the reach of his own nature.

In helping our sons affirm qualities and awarenesses that are often denied or denigrated, I would like to emphasize a further capacity of mind that can be developed, even in young children. As a teacher of what is called critical literacy, I know how empowering it is to be able to "read" between and beyond the lines of a text. While our little boys are still little, we can draw on the omniscience and authority that our position as parents confers to destabilize some of the stereotypic thinking, hierarchic dualisms, and final truths that they encounter in the world. Through questions, word play, humor, and irony, they can learn that words, meanings, and beliefs are fluid and debatable. Why should boys only play with boys, and girls with girls? Is Nowethu's God a man? Why didn't the prince ask if the young girl wanted to marry him? It could be argued that, since children at this age have a basic need for security, it is disturbing and inappropriate to promote a sense of cultural relativity in this way or to develop at this time an analytic consciousness which destabilizes certainties. I tend to see it differently. Learning such skills as young children who are participating in a loving environment (instead of waiting until the dicey years of adolescence to come upon the postmodern "abyss" of relativism), they might have a better chance as adults of living with a measure of settledness in the flux of impermanence. I don't think my role as a parent involves attempting to render the universe unproblematic or explainable. Instead I'd like to find whatever comes to hand to prepare our children to live with conviction and compassion in the beautiful, sad, suffering, changing, interconnected world of living beings, which evades all the words or metaphors or propositions we may use to interpret it.

In all this, I have assumed that parents and other care-givers are in many respects positioned as teachers. But in concluding I'd like to turn this around. Again and again and in different contexts, spiritual texts remind us to "become as little children," to rediscover the innocence of the very young. As a mother to Sophie and Sky, often exhausted and exasperated, I have found that the clear light of their open minds and loving hearts has remained a source of profound joy. As we care for our little boys and girls, doing whatever we can to prepare them for a world that none of us can really imagine, we can remember to listen to what they may have to teach us. Before and beyond the language of guns, chain saws, gnomes, or fairies, their changing nature, like ours, is continuous with the nature of plants and other animals. Their minds are made of mountains, rivers, and all beings. Their powerful innocence embodies the goodness of the earth, as big as the world. Let us clearly acknowledge this, and treasure their teaching.

Notes

1. Steve Biddhulph's influential and valuable work (for example, *Raising Boys*) is easily caricatured into an exclusive emphasis on the role of testosterone in little boys. Among others, William Pollack in *Real Boys: Rescuing Our Sons from the Myths of Boyhood* explicitly questions and problematizes what he calls "the myth about testosterone" (55–56).

2. See, for example, recent work by Leslie Swartz and Don Pinnock.

3. The most interesting text in this area that I have read is the 1996 special issue on "Mothering Sons," edited by R. Rowland and A. Thomas, in the jour-

nal *Feminism and Psychology*, in which a range of feminists describe and reflect on their experience of being mothers of sons. More recently, a M. Phil. minithesis by Rosemary Dixon extends their discussions by using a Cape Town case study to address the question: "How Do Certain South African Women Construct Masculinity for Their Sons?"

4. See the report on climate change on the Web site of the National Botanical Institute, South Africa www.nbi.ac.za./climrep.

5. For a useful discussion of this point, see *The Jo'burg Memo: Fairness in a Fragile World* (20).

6. This is a point at which to acknowledge the contribution of other people in my work on this chapter. I would like to thank my colleague Lannie Birch, who said, "Why not write about it?" when we were wondering about boys and feminism, and my husband, Michael Cope, who, as always, has helped to make this possible. Among many women friends who have discussed these questions, I would in particular like to thank Trish Lague, Dorothy Dyer, Joan De Castro, and Zannie Bock, who all wrote about their sons.

7. See, for example, the work done by Steve Biddulph, Don and Jeanne Elium, and William Pollack with regard to this.

8. For a useful discussion of dualism, see the chapter "Dualism: The Logic of Colonisation" in Val Plumwood's *Feminism and the Mastery of Nature*.

9. Numerous publications emanating from the Waldorf education system focus on this sort of emphasis. For a different take on a similar theme, see Dorothy Einon's near-classic *Creative Play*.

10. See Klein's detailed discussion of this in *Meeting the Great Bliss Queen*.

11. "There will be no equity unless the corporate driven consumer classes in North and South become capable of living well at a drastically reduced level of resource demand. Such a transformation of wealth is the central challenge of sustainability. It means to bring production and comsumption patterns up to the age of ecological constraints and equity aspirations" (Heinrich Böll Foundation 36–37).

12. See Michael Nagler's discussion of the place of a "metaphysics of interconnection" in relation to developing practices of non-violence (283).

13. I discuss aspects of this in my "Long Live the Fresh Air! Long Live! Environmental Culture in the New South Africa." The most successful, nonliterary educational text on the subject I have come across that derives from work in South African ecopolitics is Michael Cope's *Interconnectedness*.

14. See Neil Evernden's related discussion of the child's capacity for experiencing nonobjectified nature in *The Social Creation of Nature*, 107–24.

15. Orr's broad aim here is to enable students to "think broadly and . . . understand systems, connections, patterns and root causes" (*Earth in Mind* 23). See also his recommendations in *Ecological Literacy*.

Works Cited

Barker, Pat. *The Eye in the Door*. London: Penguin, 1994.
Biddulph, Steve. *Raising Boys: Why Boys Are Different—And How to Help*

Them Become Happy and Well-Balanced Men. Sydney: Finch, 1998.
Chilton Pearce, J. *Evolution's End: Claiming the Potential of Our Intelligence.* San Francisco: Harper San Francisco, 1992.
Coetzee, J. M. *The Lives of Animals.* Princeton: Princeton University Press, 1999.
Cope, Michael. "Interconnectedness." www.cope.co.za/Intercon/inter1.htm.
Dixon, R. *How Do Certain South African Women Construct Masculinity for Their Sons? An Analysis of Motherly Discourse Regarding Gendered Expectations.* M.Phil minithesis. Bellville: University of the Western Cape, 2001.
Elium, Don, and Jeanne Elium. *Raising a Son: Parents and the Making of a Healthy Man.* Hillsboro, Ore.: Beyond Words Publishing, 1992.
Enion, Dorothy. *Creative Play: Play with a Purpose from Birth to Ten Years.* London: Penguin, 1985.
Evernden, Neil. *The Social Creation of Nature.* Baltimore and London: Johns Hopkins University Press, 1992.
Gyatso, Tenzin. *A Flash of Lightning in the Dark of Night: A Guide to the Bodhisattva's Way of Life.* Translated by the Padmakara Translation Group. Boston: Shambhala, 1994.
Heinrich Böll Foundation. *The Jo'burg Memo: Fairness in a Fragile World.* wwwjoburgmemo.org.2002.
Klein, Anne Carolyn: *Meeting the Great Bliss Queen: Buddhists, Feminists, and the Art of the Self.* Boston: Beacon, 1995.
Martin, A. C. *The Durban Light Infantry.* Vol 2, *1854–1960.* Durban: Headquarter Board of the Durban Light Infantry, 1969.
Martin, Julia. "Long Live the Fresh Air! Long Live! Environmental Culture in the New South Africa." In *Literature of Nature: An International Sourcebook,* edited by P. D. Murphy, 337–43. Chicago: Fitzroy Dearborn, 1998.
Miedzin, M. *Boys Will Be Boys: Breaking the Link between Masculinity and Violence.* London: Virago, 1992.
Nagler, Michael. *Is There No Other Way? The Search for a Nonviolent Future.* Berkeley: Berkeley Hills Books, 2001.
National Botanical Institute, South Africa. Report on climate change. www.nbi.ac.za/climrep.
Orr, David. *Earth in Mind: On Education, Environment, and the Human Prospect.* Washington, DC: Island Press, 1994.
——. *Ecological Literacy: Education and the Transition to a Postmodern World.* Albany: State University of New York Press, 1992.
Pinnock, Don. *Gangs, Rituals, and Rites of Passage.* Cape Town: David Philip, 1997.
Plumwood, Val. *Feminism and the Mastery of Nature.* New York: Routledge, 1993.
Pollack, William. *Real Boys: Rescuing Our Sons from the Myths of Boyhood.* New York: Henry Holt, 1998.

Rowland, R., and A. Thomas, eds. "Mothering Sons: A Crucial Feminist Challenge." Special issue, *Feminism and Psychology* 6, no. 1 (1996).
Sister Susan. *each breath a smile*. Berkeley: Parallax Press, 2001.
Snyder, Goldin. *The Practice of the Wild: Essays by Gary Snyder*. San Francisco: North Point Press, 1990.
Swartz, Leslie. "Stemming the Tide of Violence." *Track Two* (1999): 5–9.

Nature Nurturing Fathers in a World Beyond Our Control

PATRICK D. MURPHY

It has always seemed a curious anomaly to me that so many asymmetrical verbal formations exist in American English when it comes to the creation of an embryo, that embryo's development as a viable organism, and his or her birth as a human being. A man *impregnates,* while a woman *becomes pregnant.* These phrases sound as if people still held the same beliefs as the ancient Greeks: that only the male was the progenitor of another human being, while the woman was a passive receptacle for that being's initial growth, placing each gender in distinctly different power and creative relationships to the next generation. Such inequality and distortion of contributions to life are replicated in the use of "to mother" and "to father." The former takes a lifetime but doesn't seem to begin until after the woman has given birth to a child, as if the nine months of pregnancy don't constitute a part of mothering. With the male's contribution, in contrast, every instant of his effort is credited, but apparently only up to the moment of birth.

With human beings, it would seem that in contemporary American culture a man can father a child in a few minutes—a far cry from a woman's lifetime of mothering. And yet, at the same time, if he is working on a project, such as inventing the atomic bomb, all of his work from the conception of the idea through the parturition of the object by means of a demonstration of its ability to function (or explode) is given due credit. Hence the amazing ability of the script writers for the *101 Dalmatians* movie to have the male human owner credit the male dog with the birth of so many puppies in one litter while crediting the female dog only with being too exhausted to bring the last puppy live into the world. From popular culture through scientific and historical discourses, our culture is riddled with such examples. Men are credited with creating but are not expected to nurture what they create, while women are expected to nurture what men create without being credited for participating in that creation.

As a result of such language and the thinking it both reflects and continuously regenerates, nurturing remains a concept rarely applied to men and an area of male practice inadequately studied, discussed, and promoted. Responding to the early nineties research of Sara Ruddick, which complained of the lack of analytical attention "to fathering as a kind of work," Stuart Aitken noted in 1998 that "Little is written or understood or problematized about fatherhood and domestic responsibility, whereas much more is understood about work separation, the productive capacities of men, and the power they wield over women and children" (69). Aitken goes on to paraphrase the claim of Victor Seidler that "men traditionally have great difficulty imagining the emotional space of child rearing, and their attempts are often fruitless because they have learned within a rationalist culture to deny that their emotions and feelings are a source of knowledge" (69). Charles Gaines confirms this insight in *A Family Place: A Man Returns to the Center of His Life:* "No one had ever told me that men could have access to those particular emotions and, during that time" of his children's infancy, "every day of my husbanding and fathering seemed to present me with some new impossible surprise of feeling" (33). Indeed men do remain too removed from their own emotions, not only because we devalue them but also because we fear them as an indication of weakness and vulnerability—with these two words functioning fundamentally as synonyms, even though they need not be so understood (see Aitken 69). And I can think of no other experience in our lives that generates such frightening emotions and feelings of vulnerability and, at times, helplessness and inadequacy, as child raising, where we assume we are not the experts but the inferiors. In the negotiated space and process of child raising, we have the greatest difficulty in retaining and confirming the identities and personae that we have cultivated in the public sphere.

Why does such a strong disconnection seem to exist between fathering as procreating and fathering as nurturing; between men thinking about children in terms of our roles as providers, controllers, and patriarchs, and feeling about our children as caregivers, emotional bonders, and interdependents? Carolyn Merchant and numerous other ecofeminists, such as Susan Griffin and Ariel Salleh, would tell us that historically, philosophically, and politically this disconnection is rooted fundamentally in the dualisms of mind versus body, reason versus emotions, masculine versus feminine, and culture versus nature. The chaos of child development and the situatedness and indeterminant outcomes of child raising, as well as the strong emotional reactions in us fathers that children's resistance and spontaneity elicit, contradict a rationalistic separation of mind and body, especially since these children arise from our own bodies. It is from deep within those bodies that our emotional reactions well up and overcome our fragile, mechanistic reasoning. And these emotions often intensify one such emotion, fear, in many fathers as not only do their children negate their rational, intellectual domination of the moment, but also fathers' own bodies betray us as our emotions, most dangerously anger and rage, break through our veneer of logic. An especially powerful failure of the dualisms on which we rely is reflected in the rapid resort to violence to maintain control that so many fathers elect to use against children and spouses.

These terrible—at least from the perspective of control—outbursts of emotion of any type can be particularly frightening to fathers because not only do they occur on the terrain of nurturing—traditional female territory—but also because we fathers find ourselves embodying the worst stereotypes laid on women in the man versus woman dualism: emotion, loss of control, ambivalence, self-doubt, viscerality. If fathers almost immediately become *womanly* in the face of infantile resistance, quickly surrender to all of the attributes used to define not-man, and succumb to the imperatives of their corporeality in feeling for their children, and thus let the body overrule the mind, then how on earth can male-dominated and -codified culture rule over nature? The instability, being-in-the-moment, the nonseparation of identities fostered by parenting expose the big lie of the illusion of control upon which men have built arguments for domination over nature, over women, and over their own bodily senses. No wonder we find such issues so hard to discuss, because the moment of parenting threatens us with so many realizations that few of us wish to face. But what a difference to our lives would it make for us to turn toward these contradictions of our cultivated sensibilities, to turn toward the son and the daughter, rather than toward the continuing darkness of disconnection.

Robert J. Ackerman noted in the preface to his *Silent Sons: A Book For and About Men:* "when I did seminars and lectures, consistently ninety percent of my audiences were women. Women made up eighty to ninety percent of those in dysfunctional family support groups. Where were the men, I wondered? What do sons from dysfunctional families do with their pain, their feelings, their potential? Did someone put up a 'Men Keep Out' sign, or did we put it up ourselves?" (11–12). Not only must we join Ackerman in asking, "Where were the men?" but we must continue to update the question "Where are the men?" A recently conducted five-year study of twelve thousand teenagers and their sexual activity produced considerable information about teens and their mothers but not about their fathers. Dr. Robert Blum, the study's author, is quoted as saying that "Fathers are harder to come by." In response, columnist Susan Reimer surmised, "if the fathers didn't have time to fill out the surveys and they weren't there when the interviewer arrived for the follow-up interview, then the whole 'closeness' thing isn't happening, either." Such closeness would require connection, emotional bonding, and vulnerability.

I want to consider here, then, not only how we can meet these requirements to become nurturing fathers in a culture that still fundamentally expects men to create things and people (and destroy them as well), while women care for and sustain these things and people. I also want to argue that for men such nurturing cannot take adequate hold if limited only to a consideration of how to alter and reconfigure our relationships to other people, and that it must extend also to an ecological sense of world interconnectedness and what that indicates for an altered relationship with the rest of nature. I see a way of promoting this necessary male orientation toward human nurturing grounded in a larger view of ecological nurturing. In order to do that, we need to embrace the other sides of the dualisms of culture versus nature and the masculine versus the feminine, and in particular accept our own emotions as part of our minds, our minds as part of our bodies, our bodies and those of our children as part of a natural world. Fundamentally, to undertake such an embrace means to accept interaction rather than strive to control. In order to accept how we are totally part of a series of multilayered processes and not some end product of either evolution or of culture, we will need to define ourselves as living in a world beyond our control.

In other words, all men need to become fathers in the sense of assuming ongoing responsibility for the rest of the world, not through patriarchy and the logic of domination but through heterarchy—mutually constitutive, nonhierarchical relationships. Through extending that reconstruction of men toward nurturant behavior ecologically, we can claim and promote the idea that nurturance constitutes a fundamental form of being in the world on this "symbiotic planet" (see Margulis). Accepting the responsibility of nurturing others begins with the recognition, as Lynn Margulis points out, that "All beings alive today are equally evolved.... Human similarities to other life forms are far more striking than the differences," and "Physical contact is a nonnegotiable requisite for many differing kinds of life" (5). And yet, this physical contact seems increasingly estranged rather than intimate as we continue to deny and fruitlessly fight to destroy or control even the microorganisms that both keep us alive and threaten our lives in the daily order of the world.

My father was an alcoholic who started drinking as a teenager and never stopped drinking completely until he died. He never joined AA, but if he had, he would have been confronted with the need to address this issue of "control." If people know nothing else of AA, most know that it is based on a twelve-step program. Today, millions of people in the United States, as well as many other countries, participate in a variety of twelve-step programs, all of which are based on those steps worked out originally for Alcoholics Anonymous in the 1930s. The first step of all of these programs addresses a basic orientation that men need to adopt in the search for consciously nurturant behavior. It provides a vehicle for the personal ideological reconstruction of men as lifelong, rather than one-shot, fathers, whether they pass their DNA on directly to anyone else or not. In the foreword to *Twelve Steps and Twelve Traditions*, the anonymous author observes: "Many people, nonalcoholics, report that as a result of the practice of A. A.'s Twelve Steps, they have been able to meet other difficulties of life. They think that the Twelve Steps can mean more than sobriety for problem drinkers. They see in them a way to happy and effective living for many, alcoholic or not" (15–16). And it is in the spirit of these words that I introduce the first of those twelve steps here. Step one reads, "We admitted we were powerless over alcohol—that our lives had become unmanageable" (21).

Certainly, anyone who believes that human beings are currently living in a period of ecological crisis, locally and globally, would agree that the second phrase of the first-step slogan describes the condition of existence for millions, if not billions of people. But often, those on any one of several sides of major debates about environmental threats or particular crises will see the solution as lying in exercising greater human control over other humans and, all too often, over particular aspects of the nonhuman participants of the natural world. In my experience, such has also been the case with many people who have various addictions, predilections, or obsessive desires or compulsions. If we just exercise greater control, if we just further tighten down the screws, if we just tie up all the loose ends, shore up the ramparts, more vigilantly police our children, spy more extensively on our employees, and so on, we can eliminate the variables that threaten a human-ordered world or an individually ordered microworld. For many men, that effort to exercise control can be practiced only in the microworld of the family home, with violence often the result, but for others it can be practiced in an entire nation or even the entire world, again with violence often the result.

Such efforts at control prove devastating, whether creating the dysfunctional individuals who are Robert Ackerman's primary focus or world leaders ready to incinerate millions in an effort to consolidate political power. As Gary Snyder so succinctly states in *Practice*, "It is not nature-as-chaos which threatens us, but the State's presumption that *it* has created order" (92). It is the presumption by men that they can wield any state apparatus to generate some kind of static, human-conceived order over the natural diversity and processes of life that threatens all. Hence, we must admit that we are powerless over the power of domination and that it has made our lives unmanageable.

In *The Blood of Paradise* by Stephen Goodwin, originally published in 1979 and recently reprinted, Steadman, the protagonist, decides to turn his back on inheriting the management of a development company, a position of significant power with a legacy of patriarchal control, and returns to the land, buying a small, neglected farm in rural Virginia. He takes along—and the verb here is carefully chosen—his wife, Anna, and their daughter. As with so many gentrified homesteading and return-to-the-rural-life texts, Steadman not only sees his main career as creative writing but also has a sinecure through inheritance to cushion the financial uncertainty of his move. What arise time and again in this novel and provide the strength of the plot are uncertainty, instability, and unexpected events and actions. These contingencies challenge Steadman's illusions of control and his benevolent domination of his family. Steadman can to a large extent continue to move in the direction he would like to go in terms of rural living, but he cannot control the unfolding of the path toward this lifestyle. Anna finds that her acquiescence in traveling this path cannot be maintained and is brought to crisis when she finds herself pregnant for a second time and not prepared to bear another child. She also realizes that she has let Steadman dominate her sense of reality and her sense of self, just as her twin sister had in childhood.

Rightfully so, the novel ends on a positive note of possibility but wisely does not end on a note of clear-cut resolution or guaranteed stasis. In terms of the farming that they will do and the relationship they will attempt to continue building, they both accept tentativeness into their lives. Probably evidence of the greatest likelihood of a positive resolution for the relationship can be found in the couple's relationships with their daughter. Here Goodwin portrays a man able to spend time, demonstrate patience, and provide nurturing opportunities for his daughter, on the one hand, and a woman who involves her daughter in gardening and other outdoor activities around the farm, on the other hand.

While, then, child raising receives a very positive representation in Goodwin's novel, certain ironies about control over birth and death also appear. Goodwin presents birth control, irresponsible sexual activity, and abortion as part of these individuals' common experience but points out that the birth control often fails to control. Steadman imagines that he has both freedom and certainty in his sexual relations with his wife and another woman passing through, but he is wrong. His wife becomes unexpectedly pregnant, and the result is a secret abortion. The infidelity and the abortion nearly destroy their marriage, even though both Steadman and Anna believed that they could control the outcome and impacts of these events. In particular, both of them somehow imagine that they can keep secret from the other such emotionally devastating events, with both repressing their emotions during their precipitate actions but finding themselves overcome by emotion afterward.

An additional unintended irony surfaces as a result of Goodwin's particular portrayal of birth control. In *The Blood of Paradise*, the women use the kinds of IUDs that caused irreparable harm, and in some cases death, to thousands of women in the United States. As with the side effects of the currently popular control device Depo Provera, the unintended consequences prove to be both injurious and destructive. It seems virtually a cosmic irony that men continue to expect women to assume primary responsibility not only for raising the next generation but also for controlling the population of that generation. And how often when their plans or illusions fail to maintain their control do they blame the victim or blame the technology, without assuming any responsibility?

Another novel set in approximately the same area of Virginia, *Prodigal Summer*, by Barbara Kingsolver—written twenty years after Goodwin's—takes a somewhat different approach to pregnancy. Interestingly enough, *Prodigal Summer* also contains an accidental pregnancy that results from a failure of birth control. But Kingsolver's character Deanna accepts the accident as part of the prodigality of nature and decides to see the pregnancy through to term, even though her lover has gone his own way unaware. Likewise, another female character accepts the responsibility of raising her dying sister-in-law's two children, while an old man finally learns through love to accept his estranged grandchildren. Not just this grandfather, but other male characters learn valuable lessons about symbiosis that help them in their relationships with human nature, domesticated nature, and wild nature, from lovers to sheep to coyotes. In essence, they accept the connections Kingsolver identifies at novel's end: "Solitude is a human presumption. Every quiet step is thunder to beetle life underfoot, a tug of impalpable thread on the web pulling mate to mate and predator to prey, a beginning or an end. Every choice is a world made new for the chosen" (444). Certainly Goodwin would concur with Kingsolver, for such are the realizations of Steadman in *The Blood of Paradise*, but the noticeable discrepancies between the novels deserve attention. There are good, strong women in Goodwin's novel, but they do not seem capable of articulating anything beyond an instinctive maternal caring in terms of Steadman's rural retreat self-education. In contrast, in *Prodigal Summer*, both male and female characters learn from each other, articulate ideas to each other, and engage in verbal and sensuous dialogues with each other and the interanimating world in which they grow, physically, spiritually, and intellectually. In accepting a world beyond their control, these characters act and interact in constructive, nurturant ways with the kind of responsibility that refuses the fanciful escape routes of either fatalism or autonomy, passivity or isolation—the escape routes that Steadman and Anna pursue but eventually must reject.

When I look at my history as a father devoted to promoting an ecological sensibility through my writing and teaching and at least some aspects of my lived experience, I have to admit my similarities to the real and fictional fathers I have been criticizing in this essay. Although I think I showed considerable attention to my wife in the later stages of her pregnancy and supported her during labor, I almost immediately began pulling away emotionally, quickly feeling jealousy toward our daughter, inadequacy in the face of her needs, and doubts about my identity. Like some of the fathers in Stuart Aitken's study of San Diego families, *Family Fantasies and Community Space*, after our daughter was born I immediately buried myself even further in house remodeling and the next summer signed up for more summer teaching. For several years I would pit remodeling and doing things *for* my daughter against doing things *with* my daughter. Likewise, I would use the need for devoting time to financial and career success to provide economic security for our family as an excuse for emotional distance from, and a lack of time for, family activities. Gradually I became better, or at least less worse, at these activities.

Bonnie, the woman to whom I am married, has commented on the basis of talking with other mothers that they tend to be the ones to shield and shelter the children, whether a boy or a girl, while the fathers tend to be the ones to expose them to risk and adventure. She often uses the image of fathers throwing their small children up in the air and catching them or swinging them in circles off the ground. Bonnie sees this difference as necessary for children's early development and one of the differences that suggests the benefits, if not the need, for children to have more than one parent—if not in a nuclear model then in an extended one with people who represent some of the gender differences that typically appear in society. Along these lines, I decided when to take down the baby gate to the second-story stairway in our house, even in the face of Bonnie's concern. And it was our daughter who decided whether or not she was ready to have a sky fort with an eight-foot slide. I had built the contraption and had set it up with a chicken coop–style ladder, a railed platform, and a steep slope to the slide. While Bonnie and I were discussing whether or not Mariko was ready for such an adventure, our daughter took it upon herself to climb *up* the slide and then slide down. Was that adventurousness innate or at least partly a result of our raising her with early experiences of traveling, hanging around while I worked with power tools, and put up stud and sheetrock walls, or what?

Looking back not on my actions but on my emotional reactions to our daughter, I find a lopsided reflection. Pride, anger, and jealousy more quickly come to mind than empathy, fear, and affection. But I began admitting that asymmetry a long time ago. So, which has been more important, that I expose her to adventures that her mother might choose to bypass, or that I have made a point of telling her that I love her every day? That after I have been angry, I apologize and explain that I still love her and that sometimes anger arises from love? Why do I even need to think of these actions as dichotomies?

Let me mention the areas where I have the most difficulty with my daughter. One, when she challenges me about my helping her with homework, I am more likely to become uncontrollably angry over that in the past couple of years now that she is in middle school than anything else. Why? I am an English professor, I have a Ph.D., and I know the answers. How dare she challenge my public identity and my authority, my control over the educational environment? Need I say more? Two, in fifth and now sixth grade she has become more feminine, less interested in football and basketball and being outdoors, and more interested in makeup, fashion, and shopping. This behavior doesn't anger me, but I tend to let it alienate me. The less she is interested in the kinds of activities a father would typically do with a son, the less she acts in gender-neutral ways, and the more she engages in activities that emphasize her being a girl soon to become a woman, the more distant I feel from her in that process of separation of the *male* parent from the *female* child. She is not mine; she is her own person; she is, as she always has been, beyond my control in terms of her personality, her gender, her desires and goals.

One consequence of my actions for which I would like to take partial credit has been our daughter's strong and so far lasting desire to become a veterinarian. We would like to imagine that our decision to live on the edge of town with two and a half acres of land, our appreciation for ecological diversity, our conversations about environmental problems in front of and with her, and our general eco-attitude influenced her in this direction. My own role, however, I usually believe has been quite small. I think instead that it was Bonnie's stubborn, painstaking, and prolonged care of the two cats we had who died of feline leukemia that probably propelled our daughter in that direction. Saying that, of course, makes me feel inadequate, especially since I felt jealousy toward those cats because of the attention Bonnie lavished on them. So, I asked Mariko a set of questions about why she wanted to be a vet, what events might have influenced her, and if anything her parents had done played a role. Perhaps typical of a "tweener" these days—no longer a child but not yet a teenager—she basically gave us no credit at all! Yet she clearly credited literature and games. I have decided to include her responses to the following three questions as an appendix to this essay: What events or actions in your life caused you to want to be a vet? What actions, ideas, or comments by your parents caused you to want to be a vet? Have any books or movies helped make you want to be a vet?

I had hoped for much praise from her for the various things that I think we have done, but then I realized that the way we have lived has to a large degree seemed fairly normal to her, except for our attention to organic food, which she knows is unusual among her peers. While she has not commented in her responses to our efforts to engage her with local wild nature, she does reveal a sense of an intimate contact with the nonhuman. Also, perhaps very typical of her generation, she credits books and movies with a strong influence on her thinking. And from my own view now, I see the necessity of not positing a dichotomy between direct experience and mediated experience, between practice and simulation, but rather of positing the need for more mediation and simulation to point her toward the intimacy that so many of us have lost or have had such a difficult time finding.

In this light, I come to look at Scott Russell Sanders's writing about his daughter. Scott is one of the few nature writers I know who has actually treated this relationship. Even though many others also have daughters, they are far more likely to write about their sons. Such is the case also with Scott, but he has at least made the effort to include his daughter in his nature writing. In *Writing from the Center*, she is already college age, an adult child. While she is introduced in "The Common Life," she provides only a springboard to a meditation on community. In "Voyageurs," however, she remains a focus of the essay, which treats father and daughter participation in a group canoe trip. Scott treats her fundamentally the same as he depicts his treatment of his son on similar excursions, as in *Hunting for Hope*. But he does note that within the group there is a tendency for people to try to "coddle her" (126), due more to her gender than her age. He reveals a sense of obligation to provide her with these opportunities to engage wild nature, to take risks, and to learn her own identity in relationship to the rest of the world. He provides a healthy lesson to learn about the maintaining of a certain familial intimacy while encouraging his daughter's own relational individuation. Also, he clearly sees a causal relationship between her interest in the biological sciences and the study of birds with his own commitment to promoting such experiences as this canoe trip.

In *The Force of Spirit*, Scott publishes a letter, "To Eva, On Your Marriage." In it he emphasizes her developing relationship with nature as a member of their family but frets about her coming of age in a world still riddled with sexism. And there arises a feeling of concern at letter's end about his, in a sense, turning her over to another man, in a world of men: "And while I was making the world safer for you, I would work a few changes on men as well. The prospect of your wedding has made me worry afresh about my half of the species, with our penchant for selfishness and surliness, our insecurities, our aimless hungers, and our yen for power" (133). A beautiful piece, it is followed by a similar letter to his son, and I find it significant the different sense of burden that Scott expresses here in terms of his role as a father. This role is permeated by his own position of having been a son, a burden not carried in his relationship with his daughter, such that the father-daughter relationship actually allows greater freedom for his own personal growth with, perhaps, fewer expectations about particular outcomes than he feels in his relationship with his son. This difference comes out clearly in the differing ways he depicts his relationships with his two children in *Hunting for Hope*, where he focuses on the tensions between him and his son.

While Sanders points out the cultural forces stacked against his daughter as a major concern, and rightly so, he displays an optimism about his daughter's future and the possibility of her and her husband's being able to "fashion [their] own history" (133). It is necessary to remind ourselves that no matter how cultured and acculturated they may become, our daughters remain, like all of us, a partial product of wild nature, genetically, intellectually, emotionally, and physiologically. Influence may be possible; control never. And as Mariko matures into adulthood, I have to struggle to relinquish the illusions, the ego gratification, the fear-driven desires, of exercising *power over* in order to help her through sharing *power with* her in order for her to realize her own interdependent existence. But I am afraid that I don't find much in literature to help me with this task. Even in nature writing or nature-oriented literature, I find mostly failed examples of distant, dysfunctional, and emotionally vacant fathers. Kent Haruf's *Plainsong*, for instance, depicts just such a person.

Haruf opens *Plainsong* with the main character, Tom Guthrie, clearly in crisis at work and at home: trouble as a high school teacher with disaffected and bullying students; trouble at home with a wife suffering from severe depression unable to care for her children. Tom tries to maintain some connection with the natural world through helping old friends with their cattle and involves his two young sons in the work. Yet, in that scene the sons seem to bond better with the friends than with their father, who remains throughout the novel emotionally distant from them. Haruf depicts no scenes where Tom discusses their mother's illness with them, where he shows them any significant emotional engagement, where he spends time with them away from town or ranch, even though the boys themselves are clearly interested in such adventures.

Although Tom develops into a sympathetic character, at least for male readers, he remains emotionally dysfunctional and distant. He looks to me like a prime candidate for Ackerman's workshops and a depressingly faithful rendition of far too many American fathers. Although not a fanatic for control or domination, we see him reacting with tremendous difficulty and poor judgment to the obvious loss of control he experiences in his home and in his classroom. Although his personal situation improves by novel's end, Haruf chooses not to show Tom as developing any intimacy with his sons through the course of the novel and thus not developing into a nurturing father, perhaps because Haruf remains too committed to a typically realistic portrayal of the situation to do so. One suspects similar distance in the relationship between the protagonist of Brian Kiteley's *Still Life with Insects* and his sons. And these novels focus only on a man dealing with his sons, not with daughters.

When I look for literary representations, fiction, nonfiction, poetry, of father-and-daughter relationships, I find numerous examples of daughters writing about such relationships, but little from fathers. For example, one can look at the dismal relationship portrayed in Gretchen Legler's *All the Powerful Invisible Things*, or the distant relationships depicted in Barbara Kingsolver's *Animal Dreams* and Teresa Jordan's *Riding the White Horse Home*. Jane Brox provides a portrait of a loving father of a farming family in *Here and Nowhere Else*, but one who cannot break out of a patriarchal mindset to invest his daughter with the authority she craves, and which she eventually assumes in the sequel, *A Thousand Days Like This One*.

A rare exception to the absence of fathers' depictions of father-daughter relationships in nature-oriented literature would be Chris Bohjalian's *Water Witches*. In this novel, the protagonist father initially embodies the male stereotype of dichotomous thinking. A lawyer who is also an outsider to the community, he emphasizes dispassionate logic, admissible scientific evidence, litigation, and rules, while his wife—a member of the Vermont community in which the novel is set—represents empathy and intuition through her family's history as dowsers. Initially the father sides with the ski resort industry and its expansion plans, which while boosting the economy also threaten the ecology of the area. The novel reaches its crisis when the daughter claims to have seen an endangered species on the mountain slated for expansion and the father finds himself having to believe in the *truth* of his daughter in contradistinction to the facts he has at hand. Refusing to place career, logic, and science first, the father accepts into his heart the primacy of nurturing.

To reinforce the significance of the daughter's role in the father's transformation, Bohjalian ends the novel with a coda spoken in the first-person voice of that daughter, which emphasizes not the father's education, control, or logic, but his "heart" (340).

I wish that more fathers would take the risks that Sanders and Bohjalian have taken in writing about father-daughter relationships in both nonfiction and fiction. Especially, I think, we need the nonfiction works that can match the willingness of daughters to write about their fathers. Critics also will have to begin to give this subject more attention. In the meantime, fathers would do well to read the works by daughters, to learn from the many negative and the few positive models they depict. If nothing else, both the negative and the positive fathers portrayed in the literature by both male and female writers can help me learn to relinquish the illusion of control—the same illusion that Bohjalian's father relinquishes—but I have to join that relinquishment with an engagement and enlargement of my own emotions in a nurturance based on interanimation, dialogue, and an acceptance of continuous ignorance, while all the while having to make decisions and take actions that have consequences beyond any horizon I can envision.

Appendix: Mariko's Written Responses

1. What events or actions in your life caused you to want to be a vet?

When I was younger I wanted to be someone who helps people or things. Since I didn't know what a veterinarian was I wanted to be a doctor or a nurse. Then when I was 6 I saw a veterinarian. Then I couldn't decide whether I wanted to be a veterinarian or a doctor. I knew though that I was born to love animals and I never met one that didn't like me. Since I loved animals so much I wanted to be a veterinarian. As I got older I loved animals even more. When I was born I already had a cat. When I was 1 I got another cat. I have always had animals that I loved.

2. What actions, ideas, or comments by your parents caused you to want to be a vet?

My parents never really do anything to make me want to be a veterinarian. They did do things though that helped me want to be a veterinarian after I already knew I wanted to. They bought me a book called *Emergency Vet,* which is written about real life veterinarians. Also they bought me veterinarian play equipment, which I use on stuffed animals. My parents also bought me a CD-ROM for the computer called "Emergency Vet," which is a game where you try to save animals that is created by real life veterinarians. They also helped me keep my dream strong.

3. Have any books or movies helped make you want to be a vet?

Yes, I have books and have seen movies helping me want to be a vet. I have a book called *Emergency Vet* that really inspired me. The way they (veterinarians) dedicate their lives and time to saving animals. I have seen millions of movies about vets. Two of my favorite movies are *Dr. Dolittle* and *Dr. Dolittle 2*. I thought they were so funny. I have seen another movie. I can't remember the title but it was about a vet who loved animals and dedicated her life and time to saving and helping animals. There are so many books I've read and movies I've seen I can't name them all.

Works Cited

Ackerman, Robert J. *Silent Sons: A Book For and About Men.* New York: Simon and Schuster, 1993.

Aitken, Stuart C. *Family Fantasies and Community Space.* New Brunswick, N.J.: Rutgers University Press, 1998.

Bohjalian, Chris. *Water Witches.* Hanover, N.H.: University Press of New England, 1995.

Brox, Jane. *Five Thousand Days Like This One: An American Family History.* Boston: Beacon, 2000.

———. *Here and Nowhere Else: Late Seasons of a Farm and Its Family.* Boston: Beacon, 1995.

Gaines, Charles. *A Family Place: A Man Returns to the Center of His Life.* New York: Atlantic Monthly, 1994.

Goodwin, Stephen. *The Blood of Paradise.* 1979. Charlottesville: University Press of Virginia, 2000.

Griffin, Susan. *Woman and Nature: The Roaring inside Her.* New York: Harper and Row, 1978.

Haruf, Kent. *Plainsong.* New York: Vintage, 1999.

Jordan, Teresa. *Riding the White Horse Home: A Western Family Album.* New York: Vintage, 1994.

Kingsolver, Barbara. *Animal Dreams.* 1990. New York: HarperPerennial, 1991.

———. *Prodigal Summer.* New York: HarperCollins, 2000.

Kiteley, Brian. *Still Life with Insects.* St. Paul: Graywolf, 1993.

Legler, Gretchen. *All the Powerful Invisible Things: A Sportswoman's Notebook.* Seattle: Seal Press, 1995.

Margulis, Lynn. *Symbiotic Planet: A New Look at Evolution.* New York: Basic Books, 1998.

Merchant, Carolyn. *The Death of Nature: Women, Ecology, and the Scientific Revolution.* 1980. New York: HarperCollins, 1990.

101 Dalmatians. Animation. Disney, 1961.

Reimer, Susan. "From Soccer to Succor, Moms Matter." *Orlando Sentinel,* 18 September 2002.

Salleh, Ariel. *Ecofeminism as Politics: Nature, Marx, and the Postmodern.* London: Zed, 1997.

Sanders, Scott Russell. *The Force of Spirit.* Boston: Beacon, 2000.

---. *Hunting for Hope: A Father's Journeys.* Boston: Beacon, 1998.
---. *Writing from the Center.* Bloomington: Indiana University Press, 1995.
Snyder, Gary. *The Practice of the Wild.* San Francisco: North Point Press, 1990.
Twelve Steps and Twelve Traditions. New York: Alcoholics Anonymous Worldwide Services, 1996.

when tillage begins
A Family Portrait

JIM HEYNEN

My father was the alpha male in our neighborhood. While neighbors were still rubbing sleepers from their eyes, the lights in our barn announced the fact that Hilbert's hands were already warm on the cow udders. In early spring, the rhythmic clicking of his corn planter told other farmers that the soil was ready for planting. Both young and old men tried to follow my father's example as he stacked bales until sunset, sweating so profusely that his shirt turned the color of blue rain clouds. He was the model to live up to. He led, the others followed. As a child, I listened to neighbors seek his advice. And expert services. They'd call Hilbert before they'd call the veterinarian. Castrating bulls, vaccinating pigs, treating steers for hoof rot—he was ready for any task.

My father was not a boastful man, but I remember the day two younger men challenged his status. It started as a game of tug-of-war after the fieldwork was done. Two against two, three against three, three against four. Lots of laughter and joking and sizing each other up. Then the challenge: two younger but smaller men said they'd like to take on Hilbert. Without hesitating, my father gripped the end of the heavy rope, dug his heels into the Iowa loam, and dragged the upstarts ten feet across the field. I think he caught them off-guard, pulling before they were ready—but they didn't ask for a re-match.

My father's defeat of the young men may have taught them a lesson, but he didn't mock them. He may have been top dog, but he didn't gratuitously flaunt what he had. In the stiff-lipped Dutch Calvinist community where we lived, he understood that modesty, or at least the appearance of modesty, was the hallmark of real strength. False bravado was for the reckless years of adolescence. Peacocks. Bull dogs. Stallions. For a grown man to strut in the manner of those creatures would be worse than phoniness: it hinted of ungodliness. Gross puffery. Vile pretense. Satanic slitheriness. Real men like my father moved with confidence but not swagger. I wouldn't say he embodied that enviable quality Castiglione described in *The Courtier* in the fourteenth century: *sprezzatura*, the ability to do the most difficult of knightly tasks with apparent ease and grace. No, my father was not graceful with his big hands and substantial waist, but he was sure-footed and clear in his intentions and actions. Maybe he was more like the horse that the knight rode, and the knight that rode my father was the ideal of manly proficiency in a world where nature was tamed, harnessed, and subdued into providing sustenance for his family. He was not pretty, he was not cool, but he was efficient. He got done what needed doing.

That decisive but unassuming strength was given a life-and-death test one day when he was in his fifties. A young man who owned a custom feed grinder mounted onto a big truck came to our farm to grind a few wagonloads of corn for the pigs. The large commercial grinder had a huge diesel engine that, at full speed, sounded like an airplane. With the engine roaring, the young man's coat caught in a moving gear and in a flash wound up around his neck. My father heard the rat-a-tat-tat of the slip-clutch on the gear shaft and saw what had happened. By the time the machine was stopped, the coat was like a tightly braided noose around the young man's neck and his face was a bulging blue. While the ambulance was en route, my father deftly but forcefully cut through the noose, pulled the unconscious body away from the machine and blew air into the lungs. The ambulance driver looked at the blue face and said, "He's dead."

"No, he isn't," my father responded. "Are you going to work on him or must I?"

A week later some of our family went to the hospital to visit the recovering young man with his swollen neck. There were no celebrations, no elaborate thank-yous. Gratitude for noble action was soft-spoken in that community. If my father had given his own accounting, it would have been as simple as, "Well, you couldn't just stand there and look."

Although Grant Wood's paintings of rural Iowa people are sometimes tinged with irony, my father might be seen as one of Wood's Lincolnesque idealized men, acting with noble certainty in all situations while remaining focused on the primary task of giving shape to the untamed landscape. In the large piece that hangs in New York's Whitney Museum, *Study for Breaking the Prairie,* a prominent part of the painting is a quotation from Daniel Webster that practically dictates how viewers should respond:

> When tillage begins, other arts follow. The farmers therefore are the founders of human civilization.

If the pioneers who broke the prairie lands of Iowa in the 1840s were engaged in an artistic act, then my father's work on that huge canvas of a landscape was the advanced creative work: putting up fences and building barns, draining wetlands and digging ditches, killing weeds and varmints, and raising livestock for milk, eggs, and meat. I doubt that my father had any such grandiose conception of his place in the world during those many years that he struggled to put food on the table and a roof over our heads. He would laugh at any notion of himself as an artist. I would too.

Although my recollections of my father as a model of leadership and strength in a rural Iowa community are honest and accurate, I know that what is left out of the quick-sketch portrait is the bigger picture. The bigger picture shows a chilly detachment from both the land and the animals. When I think back to my youth on an Iowa farm in the 1950s, what I remember most clearly is not the artistry of farming but the matter-of-fact brutality of it all.

I have these images of my father:

He is castrating squealing pigs while my brother and I hold them.

He is beating a horse that has refused to go into the barn when directed.

He is cinching a rope around the legs of a kicking heifer whose head is locked in a stanchion while she struggles against her first milking and the rope rips the hair from her hind legs.

He is stamping rats to death as they scurry out from under the woodpile.

He is prodding squealing pigs with an electric shocker as they are loaded onto a truck for market.

He is sawing the horns off full-grown steers with no anesthetic as their eyes roll back in their heads.

One of the most painful memories is a scene of violence I didn't even witness. When we returned home from church one Sunday morning, we found our young dog Whitey in the chicken yard with a dead pullet in his mouth. Within an hour a shotgun blast ended the dog's life. My father did make certain that I as a boy did not see the actual killing of this wayward pet, but there was no hesitation. We didn't talk about ways of restraining or training the dog. It was a black and white issue, as free from ambiguity as the sheep-and-goats Calvinistic religion to which the entire community subscribed.

My brother and I had 4-H calves that we became very fond of as we fed and groomed them for showing at the county fair. Some of the designs we managed to currycomb onto their flanks did have a fleeting kind of artistry to them, about as permanent as snow-angels, but my father reminded us that what we were really doing was learning how to raise animals for profit. He pointed to the bottom line, and 4-H requirements were such that we had to show the cost of every ounce of grain we fed these animals. After we groomed and showed them at the fair, they were auctioned off for slaughter. We were supposed to be proud of how much money they brought. We were supposed to put aside our feelings toward animals that had become our pets. Our father taught us to let go.

Much of that fatherly education could be seen as good animal husbandry. You cared for animals so that they would care for you—by providing money and food. Animals may have been warm-blooded fellow inhabitants on the earth, but they existed for our well-being. They were commodities.

Even as I recall the detachment toward and the pain of the animals, I acknowledge that my father was a model of deportment. He was a religious man who moved through the world with a clear mission to provide for his family and to work in harmony with what he believed was God's will. Other men, and certainly we children and my mother, looked up to him with hardly a doubt as to the rightness of his actions and decision making.

My father in his world was representative of what happened across America as land was settled and developed for agriculture. But for me it was hardly a Grant Wood world. Rather than artistry, it was defilement, shortsighted exploitation that translated into workaday violence. Of course, the detached acts of defilement have accelerated since my father's time: Iowa has increasingly become a corporate world of huge cattle feedlots and hog confinements. Much of the countryside carries such a stench that it is hard to imagine anyone pausing to admire any beauty of land or animal. Not only is it not Grant Wood's world, it wouldn't fit Jefferson's romantic vision of a land populated by small and self-sufficient farms.

Still, I am not sure what the alternatives might have been, and I am left with a series of troubling questions: Were the Iowa settlers violent when they turned the prairie grass into farmland? That had been done by a generation or two before my father.

Were they violent when they put up fences for their cattle? That too had been done before my father got there, but he certainly continued in the tradition of building good fences, if for no other reason than to have good neighbors.

Were they violent when they started coaxing larger and larger corn yields by applying richer and heavier fertilizers? I was certainly part of this process, working as a teenager in the commercial application of liquid nitrogen to cornfields.

Were they violent when they applied pesticides against Canadian thistles, leafy spurge, cockle burrs, milkweeds, corn borers? Ah, here is where my boyhood memories intensify. The 2,4-D, the DDT, the aerial spraying of cornfields for corn borers—all smells that linger clearly in my olfactory memory and some of whose contents no doubt still linger in my blood and probably in the blood of my children and (a slight dread I carry with me) may manifest themselves in the bodies of my grandchildren and great-grandchildren.

My older brother owns and works the family farm today. I am grateful that it does not resemble the huge corporate operations, but it does show the history of what has happened to the land. When I visit, I often walk across the fields where my father spent most of his waking hours. I like to pause along the deep ditch that replaced the meandering creek of our childhood. This is where slew grass once grew in about a twenty-acre area that could accurately be called wetlands. I remember the laying of the tile that would drain the wetlands and empty the excess moisture into the new deep ditch that was carved into the land. This was progress. At the time, my father's decision seemed like such a logical thing to do: drain the land and straighten and deepen the creek so that straight rows of corn could be grown in place of the relatively useless slew grass. It's not as if my father was the only farmer to "improve" his land in this manner. But he, as usual, set a neighborhood precedent, and when others saw how productive that converted slew grass was when it was turned into corn land, many others followed suit.

What was lost in this seemingly bloodless land improvement? Here's a starter list: mud turtles, skunks, badgers, civet cats, garter snakes, weasels, crawdads, field mice, chicken hawks, foxes, rabbits, fireflies, toads, crickets, June bugs, redwing blackbirds, bullheads, pheasants, swamp ghosts (or will-of-the-wisps, which still live as live creatures in my memory), and a multitude of shiny and multicolored beetles and other insects that were gone before I learned their names.

I often wonder now if my father had any sense of what he was doing as he moved through the world with his powerful efficiency. As a child, I felt the animals' pain. I think I also felt the land's pain. Much as I respected my father, and still do, I don't think he could feel nature's pain—until he was an older man—and this was a time of great transformation for him.

I remember first seeing a sign of a man who empathized with the creatures on his farm when he was getting older. He still had some old-fashioned farrowing pens for his sows—the really old-fashioned kind that provided an eight-by-eight foot, boarded-off space where he'd throw straw so the expectant sow could design her own private nursery. Design is exactly what the sow did. She would take mouthfuls of straw and use her snout to shape a big, egg-shaped nest of straw. Then she'd flop down into the middle of it and wait for the contractions to begin. The spacious farrowing pen gave the occupants plenty of room but not much safety. Too often a sow would step on her young or lie down and smother them before they could scurry away. This is exactly what happened one morning when I was visiting the home farm and accompanied my father into the hog house for the usual morning work of feeding and cleaning. The sow rose to the sound of our coming, but a tiny Chester White pig lay lifeless in the straw. My father stepped into the pen, disgruntled at the sight of another dead pig, and flung the limp form onto the manure pile. When it landed with a *thunk*, the pig emitted a small grunt. A little cough.

My father retrieved the pig and with his big hands gave the rib cage a gentle squeeze. The small throat opened and gasped for air. It was still alive but couldn't breathe without his help. For the next crucial minute, I watched his large hand tap on the rib cage while he blew little bursts of air into the pig's mouth and squeezed rhythmically with his other hand. This was not the brute force of a man defeating young challengers at tug-of-war. This was not even the quick and forceful hands that saved the life of the strangled young man who came to grind our corn. It was a gentle and earnest caring that I had never witnessed in him before. I can still hear the *tap-tap-tap* on that little pig's rib cage. I can almost feel it. With those big gentle hands, he coaxed the pig back to life.

As usual, there was no fanfare in this rescue mission. When the pig squirmed and let out a little-pig squeal, he stroked its forehead, smiled, and said, "You'd better not try that again." He placed the pig back into the pen and, with that, went back to work.

My parents never traveled more than about a hundred miles from the farm where they lived. When they were in their late sixties, I volunteered to take them on the trip of their lifetime—the Black Hills, Yellowstone, and Glacier National Park, before arcing up into Canada and reentering the United States in Washington State, where we have relatives. As we were traveling, I was surprised by my father's exact and insightful observations. He noticed colors, subtle changes in landscape, grazing antelope that didn't catch my eye. He was like a child, delighting in every nuance of the changing scenery. I was humbled by the clarity of his vision, often a brilliant innocence in the way he contrasted what he knew so well from years of farming in Iowa with what he was seeing for the first time. After several hours of driving through the Canadian Rockies, with their seemingly endless expanses of evergreen-covered slopes, my father said from the back seat, in a voice that hovered between awe and sadness, "You know, there may be more trees in this world than stalks of corn."

In that moment, which was comic only for me, I realized how deeply he had internalized the world from which he seemed to be so detached all those years when I assumed he regarded his animals and land as mere commodities. They were more than commodities, they were the very makeup of his being, a set of lenses through which he could more deeply appreciate the richness of the larger world.

JIM HEYNEN

Some years later, when he was in his late seventies, he would drive from his retirement home in town and help on the farm. One spring he was driving a large tractor with a plow behind it. He'd clutch the tractor and pause, then start up again to move a few inches. He kept inching along in a manner that he might have called "wasting time" when I was a teenager and expected to get the fieldwork done as quickly as possible. Then the reason why he was making the many stops became apparent. A baby cottontail rabbit was hopping along in the furrow just a few feet ahead of the tractor tire. In his younger days, I'm sure my father would have had as much respect for the cottontail as Beatrix Potter's Mr. McGregor had for Peter Rabbit invading his garden. On the farm where these nibblers destroyed crops, the only good rabbit was a dead rabbit. Clearly, as he aged, my father's attitude toward useless creatures was changing. At that moment, he had no need to subdue the earth or have dominion over its little parasitelike inhabitants.

The new tenderness that emerged as my father aged was nothing short of a marvel. In his eighties, he'd walk the cattle feedlots that were now operated by my brother. He got to know the animals individually and would talk to his favorites. You might say he came to pay them a visit. He'd hold out his hand for no other reason than to connect with them as fellow creatures. He never talked about how many valuable pounds per day they might be gaining. He had moved beyond the bottom line to what seemed like a genuine concern for their well-being.

In his final year of life, I'd visit him in the nursing home. I never saw him reading his Bible. When concerned religious people asked him about that fact, he said, "Oh, I've read that plenty of times. I know what's in it." Instead of studying the Scriptures, he liked to sit near the entranceway where there was a large aviary with dozens of birds with various small nesting compartments. He'd watch them for hours. When some of the birds were singing, he was more interested in them than he was in talking with me.

I've tried to understand the change that came over Hilbert in his later years. It was more than the fact that his money worries were over. It was as profound as coming to think differently about God. During those years of mandated service, doing—as he would pray—God's will, he was a person driven by an idea of doing right in the world. He was a person for whom religion began in his head. Maybe that's how he farmed too—from an idea of how to do things for the best monetary results. In his later years, it was as if he started experiencing God in the creation and started feeling a greater reverence for all that breathed around him. Perhaps this is what his years of tilling were all about, to get to this moment.

And what if this transformation had come much earlier in his life?

Caught on that grindstone of responsibility to his family, brainwashed by all the signals of what it means to *progress*, could he have clung to a more balanced and—dare I use the word?—*holistic* approach to farming: crop rotations, diversified livestock, a farm with wetlands and pasture as well as corn? Only if he could have seen into the future, I suspect.

By the time my father was in his eighties, his sensibility was totally out of sync with agricultural "progress." He saw the evils of corporate takeover of small farms. He saw what was being lost by turning the landscape into a factory. If he had had his empathic sensibility as a younger man, would he have spared the slew grass? Would he have continued crop rotation over heavily fertilizing the land so that nitrogen-depleting corn could be planted year after year on the same soil? Would he have preserved the diversified farm that in his younger days meant some chickens, some milk cows, some pigs, some feeder calves—instead of the current trend to focus on one thing such as fattening cattle in large feedlots?

If he had been able to foresee what was going to happen, I am convinced that the Hilbert I knew in his thirties and forties would have been much more like the Hilbert I knew in his eighties. My perspective is, of course, the most privileged. The few farmers I can remember who did resist progress were the first to be eaten up by progress—forced to sell and move to town and work for someone else. My father's approach did result in economic rewards, many of which benefited me. Part of what he was doing in those years was making certain I could become educated and have the freedom to do what I am doing right now. Aye, there's the rub.

Husbands and Nature Lovers

LILACE MELLIN GUIGNARD

Usually it's bad when a relationship goes on the rocks, but climbing—in addition to paddling, hiking, camping, and mountain biking—grounds my marriage. Jimmy was a climbing and surfing Eagle Scout–redneck when we met in western North Carolina while I was a kayaking and raft-guiding feminist-hippie. Throughout our early shared outdoor experiences, I could tell Jimmy (correctly pronounced *Jeemy*, like the Carolina wren's call) was watching me, gauging my actions, reactions, and decisions. Quickly he distinguished himself as one of the best of the good men in the outdoor recreation subculture, as he directed his gaze more toward my belay technique than toward my tight black shorts. At least that's what it felt like. As his buddy Sean climbed, attached to a rope looped at my waist through a device designed to help me catch him if he fell, I appreciated the watchful eye that was not in the least condescending or judgmental. Neither did Jimmy try to take over and do it himself.

One of the first things people notice about Jimmy—especially if he's climbing without his shirt—is that he's covered with hair. My girlfriend who introduced us had warned me. Even Jimmy's brother teases him that the only place Jimmy doesn't have hair is his knuckles because they scrape the ground when he walks. Yet he's not some caveman type at all, focusing only on his challenges and exploits, treating each climb like a mastodon hunt. He and his buddies decide together what routes to do and in what order they'll climb using a sharing, rather than competitive, model of communication:

"John, you should lead this one."

"I led the last route yesterday, Sean. Why don't you go ahead? I'll clean gear."

Climbing is about partnership—shouldering responsibility while simultaneously relying on another. Jimmy's taught me this. Though he is competitive, he never sees the challenge as conquering the rock. Even as I start to climb harder than he does on certain types of routes, he just grins below his cap brim and says to whoever's near: "That's my wife!" We have pretty much the same conversation when one of us has just followed, or top-roped, a climb the other one led (a riskier job):

"Watch me!"

"You've got it. Breathe."

"Sonuva... Whew!"

"Nicely done."

"You got me?"

"I've got you."

"Okay, lower. Damn, that was scary. Great lead!"

"Thanks, but you made the crux look better."

Usually it goes just this smoothly, with us alternating roles and lines all day, safely lowering each other to the ground. But sometimes the route is harder than it looks, riskier due to loose rock or choss, or a big storm comes up fast when we're several pitches (rope-lengths) above less lightning-prone zones. Then our interaction is not as casual but just as precise. It's during these times that we learn how the other responds when stressed and develop techniques for working together under pressure. Jimmy has learned I won't swoon, get paralyzed with fear, or put the pressure on him to save me just because he's the man. I've found out he won't blame me for the situation, try to control everything, or take off in some every-man-for-himself fashion. Even though it was years before we had a heated fight, because of our interaction outdoors I trusted—even as his fist went through the drywall—that he wouldn't unleash his frustration on me or take off with no word the way several friends said their lovers did. And I was right.

Jimmy is certainly not the only man out roaming the rocks, waters, and woods who isn't propelled by a man-versus-nature worldview. He's not the only man who enjoys sharing (really sharing) his outdoor activities with women. I have found a large number of such men in the subcultures of paddling and climbing, men who encourage women to realize their personal and muscular potential. I watch as more and more young guys learn to climb among and even from women, and these teenagers often share their insecurities—about climbing *and* dating—with their female peers. I hope this is a real sign of future change. Sadly, the brawny "Bluto in Paradise" type, as one friend labels the testosterone bullies (comparing their packs to see whose is larger and their climbs to see whose is harder), is still in the majority. So much so that single girlfriends who join us climbing on California granite or hiking in the high desert often wistfully wonder aloud as we leave Jimmy to find a modest spot to pee: Where can I find one like him?

Where indeed? For me it was the South, the Bible Belt, that supposed bastion of the white male and his cohorts—racism and sexism. The realm of NASCAR and motocross and monster truck rallies, where masculinity is often linked to mechanical toys that eat up natural resources quicker than the family can put away liver mush and biscuits at Memaw's Sunday breakfast. Our last trip home coincided with several races, and we caught some of them while visiting relatives. At Nanny's we saw the beginning of the Indy 500, exchanging bits of news while her ears (she's legally blind) were trained on her boys. And at the folks' house we watched an exhausted Mark Martin win the Coca-Cola 600. Coincidentally, this was the Viagra car, ensuring in myriad ways that the South shall rise again.

The first person in his family to pursue a doctorate, Jimmy currently studies Victorian literature and ecocriticism at the University of Nevada at Reno. Here's his alternative reading of NASCAR: "Don't think these guys aren't real athletes just because there's a motor involved. It takes skill and strength and freaking *amazing* coordination to concentrate for five hundred miles going two hundred miles an hour." Jimmy points emphatically at the TV screen. "These guys have more in common with extreme sports than organized ones like football—the risk is huge. I know some folks would say NASCAR is just getting in a car and turning left, but they don't appreciate the physical and mental requirements of good driving." Coming from a family of long-distance truckers, Jimmy knows what he's talking about when it comes to driving. Even though I'll never understand the pull, this is part of Jimmy's heritage and he respects it. But with his free time he prefers to put his body directly in contact and communication with elements other than steel, to be elevated and humbled at the same time by engaging with forces greater than him. His legs and arms are his pistons. When Jimmy says a climb or river kicked his ass, he shakes his head while grinning the same way he does when a buddy bests him.

It wasn't until recently that I seriously considered what forces shaped Jimmy's relationship to nature. How has his masculinity been affected by cultural attitudes toward nature and how has he avoided being twisted in some of the more negative ways? The list of generalizations about masculine versus feminine ways of "being in relationship to nature" includes: controlling versus caring, mastery versus harmony, arrogance versus humility, and acquisitiveness versus appreciation (Anderson xvii). These approaches emphasizing dominance over communion are inherently violent. In addition, separateness and self-reliance became masculine traits forged in the Teddy Roosevelt and Frederick Jackson Turner school of wilderness. I agree with Susan Faludi that "the man controlling his environment is today the prevailing American image of masculinity" and that "a man is expected to prove himself not by being part of society but by being untouched by it" (10). By this definition, one of the qualities of manliness is to actually *be* a wilderness. But now civilization is finding its way into the wilderness in the form of women, and a man often doesn't know how to react—should he be the heroically detached outdoorsman or the protective, family-oriented provider? Our culture has made it hard for men and women to be together far away from home and hearth. This is reflected in the riddle:

> Q: If a man speaks in the forest, and his wife's not there to hear him, is he still wrong?
> A: Why do you think men go to the forest!

So, when looking for a life partner, what type of man does a nature lover look for? It's been my experience that many men have unexamined assumptions about women who choose to lead their lives in more natural, lonely, or rugged environments. If the pursuit is not linked to something domestic like ranching and farming, the female raft guide or Wilderness Ranger feels the unspoken question in the men trying to sniff out her reason for being there. The two main options our culture has historically made available are: (1) she's a loose woman who has crossed into the wife-free zone of wilderness where good women just don't wander, or (2) she's really a man, a butch dyke, a nonwoman. While I don't believe women are biologically or mystically closer to nature than men, I do believe that a man's treatment and consideration of nature is inextricably linked with how he interacts with women—especially a wife or lover.

Men who operate, consciously or unconsciously, under the two above assumptions are still embedded in the psychosexual pastoral impulse that Annette Kolodny deconstructs in *The Lay of the Land* (1975). They are stuck treating the land as woman, a virgin to be conquered or plowed. "The quality of [the American] experience is variously expressed through an entire range of images . . . including eroticism, penetration, raping, embrace, enclosure, and nurture, to cite only a few," states Kolodny (150). Many women find that this collapsing of women and nature (and sexualizing both) makes interactions with such men unpleasant—especially outdoors. And, once again, those interactions affect how such men treat nature. Before Kolodny, Paul Shepard warned that "the relationships between men and women partly determine how people use their environment" (109). But centuries-old assumptions take generations to fade away. So, as a wife, an environmentalist, and a feminist, I ask myself to what serendipity do I owe the fact that Jimmy both interacts easily with women in nature and interacts with nature in a way that makes him easy for women to be around outdoors?

First, let's look at his childhood in the South, specifically in North Carolina (except for a few years in Mississippi) with summers spent at his grandparents' home just over an hour away from the piedmont, thousands of feet higher in the Appalachians. His family goes way back in the town of Banner Elk, with many still living on parts of the original four-hundred-acre farm. There Papa always grew potatoes, greens, beets, corn —everything but okra, which needs the hot flatlands. There Jimmy's daddy learned to garden. There Jimmy picked apples and blueberries, dove in the creek to escape yellow jackets, and climbed the hemlocks and birches. There he learned to love not just nature, but a place. He learned that husbandry is inherently domestic.

Jimmy seems to have benefited from the southern culture that sends what might be considered mixed messages about men's and women's roles and strengths. It's the land of the Good Ol' Boys network but also of the Steel Magnolias—strong women who, while the fathers and husbands may be more visible in public, are the ones anchoring families through the day-to-day tough times. Depending on the situation and individuals, this arrangement can be an unfair division of labor or a highly successful interdependent economy. Jimmy's daddy hauled produce and then hardware cross-country in a cab-over Kenworth rig. Jimmy's mom worked full-time in the school office while also keeping a warm, fun house at which all his and his brother's friends would hang out. Jimmy, like his daddy, grew up being passed from driver to driver as a tyke, and taking the wheel a few times as a teen—slipping into the driver's seat while his daddy slipped out, doing seventy down the highway. On these trips his daddy taught him to notice the red-tailed hawks on the snags and poles and, at dusk and dawn, to spot deer just inside the trees lining the highway. When Jimmy's daddy was home, he was one of the local scoutmasters.

In the early 1900s, after the frontier was gone for good, "the Boy Scout movement was one answer" to saving our future men from being over-civilized now that there was no more wilderness to conquer (Nash 147). But the South has never been as preoccupied with wilderness as the rest of the nation. We do have laurel hells and tangles of rhododendron and doghobble in Appalachian hollers that once were the demise of hunters, and those like Eric Rudolph who know the mountains and caves can still hide from even the FBI, yet these aren't usually considered wildernesses in our national imagination. The South has a pastoral relationship to nature and has historically depended upon a settled agrarian social organization rather than an ethic of rugged individualism (Kolodny 136–37). Jimmy's male role models weren't out conquering wilderness alone— they were gardening to feed their families. In this context it makes sense that Jimmy's tasks as he worked his way up to Eagle Scout had more to do with community service than woodcraft. Sure, they'd go backpacking, but just as often they'd be repainting church parking lots. And his Order of the Arrow test was a mixture of both hard group labor and a white boy's vision quest. On little more than a boiled egg, the chosen scouts repaired steep, eroded trails all day and then spent the night alone in the woods with only a sleeping bag and water.

So from an early age Jimmy experienced getting out in nature as a means of developing community as well as getting away from people. Even now when we go on a climbing trip or hike, he's always talking with folks we pass. Jimmy Never-Met-A-Stranger Guignard, our friends tease, and it's true. Sometimes it drives me absolutely nuts because *I'm* the one who wants to get away from people, to shed society even though, thanks to Jimmy, we've made many friends this way. As he slows his stride to match some middle-aged hiker's, often I'll speed down the trail—the loner growling savagely about how crowded it's become, disappearing into the shadowy forest.

But this community ethic is only one part of Jimmy's relationship with nature. When he was growing from the blondish good boy into the tall, hairy brunet man who could drink cases of beer each night with his buddies, Jimmy went to college at Appalachian State University near his grandparents. There, a Yankee friend got him into rock climbing. Neither of them knew what they were doing at first, but they borrowed gear, Sean checked climbing books out of the library, and Jimmy took him into the mountains at the foot of which he'd grown up playing. From the stories they tell, they gained their skill slowly and cautiously, each looking out for the other and depending on each other's opinion as they taught themselves how to assess new situations. When I ask Jimmy why they weren't gonzo risk-takers like stereotypical teenage males, he replies, "'Cause we were chicken . . . ?" and laughs. "Seriously, getting hurt didn't mean you were tough—it just kept you from going outside. We knew we didn't know all the things that could go wrong, so we backed up everything. If slinging one tree was a good belay anchor, we'd sling three."

Scholars say the outdoor recreation movement grew out of the need to turn a civilized boy "from a weakling into a man," and that's what happened to Jimmy, at least physically (Evans, quoted in Nash 153). In videos of him back then he's tall and lean, and his belly is a washboard you could play rhythm on in an old-timey band. The hair dappling his shoulders and lats casts forest shadows across his ripped muscles. He wears large tortoise shell glasses popular in the seventies, a five-day-old beard grows like kudzu, and his mane of brown hair is untamed despite the bandanna. As he turns toward the camera describing the hardest part of the route he just did, Jimmy points and thickly drawls: "Them's the tweakers, dude."

I watch this and can hardly believe the wild man from which my husband evolved.

Jimmy's love of climbing, and the friendships with Sean and others that went along with it, was a door into a more complex understanding of himself and the idea of "nature." As a mountain biker he learned the difference between national forests and national parks in general and wilderness areas where bikes aren't allowed. As a climber he appreciated that there was "nature" constructed so that the buzz of dirt bikes didn't intrude on the conversations between him, the pines, and the cliff. Even if he didn't question why till later, he began realizing there were certain types of "nature" he preferred over others, and certain types of physical activities that made him feel alive and connected in ways that heavily mechanical sports didn't.

Jimmy didn't become suddenly environmentally conscious, of course. He was still maturing in general, still majoring in partying and climbing. But I think it's significant that during this time in his late teens and early twenties, Jimmy found the rocks and mountains to be places that helped him escape society enough to find out who and what he was, and the guys he escaped with to be good companions on the journey into manhood. What they discovered separately and together was knowledge they carried back into town. I imagine him out in the chill mornings before class, his ankles damp from the dew on the galax as he moved from boulder to boulder on Grandfather Mountain, testing his strength and stamina against his mentors'—the rocks. He'd never best them, but he could join them if he could find a way to the top. Back in class he'd read books and take tests, but out bouldering he'd feel all that he learned rush through his body before his brain could process it fully. These chunks of granite were as much his teachers as were his grandparents and parents and professors.

Ironically, Jimmy's environmental awareness bloomed during his six years working construction. After college, at twenty-two, he became job superintendent for a general contractor working mostly government gigs such as bringing a hospital up to fire code and landscaping HUD housing developments in Charlotte. Today Jimmy loves to tell fellow academics and environmentalists about the job dredging a reservoir and raising the dam and spillway when he got to drive a bulldozer because the subcontractors were a man short. "Eddie put me up in the cab of the big TD-15, showed me how to raise and lower the bucket, then sent me over to the borrow area to push down trees." The borrow area is where they would get the dirt to raise the spillway. "I mean this dozer was *huge,* and I had to drive the heavy metal beast across a narrow dam. Then I rode right up the trunk of one of the oaks till it fell over. After a few more, one of the regular guys on the crew suggested I come at the trees from the other side so they'd all be facing toward the road for when we had to pull 'em out." Jimmy counsels that it's important not to underestimate the thrill of power and control one gets from operating big machinery. When we drive through road construction, he still stares at bulldozers wistfully.

Several things happened during those years that made a lasting impression. One was when a mega-grocery store went in near his grandparents' home. "They cut the whole hillside out. Hell, one of my favorite trails had been there." Development everywhere became a big concern of Jimmy's as he saw the geography of small Appalachian towns change when shopping centers and planned communities took the place of individual stores and homes that could adapt to the land. But the issue isn't simple, Jimmy stresses. "I'm not angry at the local boys who got the contract to level that hill. I was never angry at those guys. I'd rather it be them if somebody's got to do it. Who I'm mad at is the big corporations that don't know firsthand the place they're putting these buildings. It's just some *For Sale* spot on the map."

After working construction four long days a week, Jimmy would drive from D.C. or West Virginia or South Carolina back home. For a while he mostly mountain-biked and road-raced on weekends. When he started climbing again he was stunned by the impact on one climbing area right off the Blue Ridge Parkway. Over the last couple of years Grandfather Mountain had been closing down public access because the owners were afraid of being sued if climbers got hurt. So the crowds all flocked to Ship Rock, on national park land, and by the time Jimmy returned the trails were eroding and litter could be seen peeking out of the galax and mountain laurel. In Boy Scouts they'd focused a lot on the problem of pollution, but from working construction, Jimmy started to put what was deeply personal—that spot on earth—into a larger environmental context of development and population. His involvement with bulldozing healthy trees and putting up condominiums makes it so that, even now, he cannot talk about environmental issues as if he's above or outside of the economies that are trapped in the cycle of creating short-term gain while ignoring long-term loss.

I met Jimmy years after he quit construction, when he was beginning graduate work in English but looking for a way to combine his studies with his love of the physical environment outside of books and classroom walls. By this time he stood six-foot-two, wore contacts, was barrel-chested and clean shaven, with shoulder-length surfer hair often worn in a ponytail showing off his Paleolithic forehead to good advantage—he was the epitome of physical manliness. Best of all, he had a well-developed . . . community ethic and environmental consciousness.

LILACE MELLIN GUIGNARD

In addition, Jimmy was already one of those guys sought by women of all ages for advice and an empathetic shoulder. "He's a real arm-patter," Ruth, an eighty-three-year-old woman, used to tell me when Jimmy and I started dating. It's hard to figure why he was still single unless he hadn't met anyone who'd join him mountain biking and climbing instead of resenting the time he spent doing those things. To my amazement, I found that Jimmy had developed all the stereotypical southern traits of hospitality and courtesy without any of the condescending "little lady" attitude toward women. It must be because no males in his immediate family have that attitude either, though I can't say why. He wooed me by frying some of Papa's potatoes in butter till golden and a little crispy, standing barefoot in my kitchen. He says I passed the first test when I let him open the pick-up door for me.

There were other tests we had for each other the following weeks and months, ones having to do with how we could work and play with someone else, not completely giving control over to the other or demanding complete control over the other. I was especially interested in how he handled my varying levels of confidence and fear in our shared recreations. Mountain biking was the most challenging activity for us as a couple. Jimmy was so good and knew so many riders who were even better than he was, that his perspective was skewed. When he'd tell me he'd picked out a moderate trail, I'd come back thinking, "like Hell." But he didn't scoff as I tried to explain how I felt and what I needed in order to gain skill without scaring myself witless or walking my bike for miles. This, in turn, forced me to consider what I did need, rather than putting the responsibility on him to *know* what I needed and plan appropriate outings.

As I learned more and more about climbing by going cragging with him and his friends, he was starting to kayak with me and my friends. Watching Jimmy approach whitewater with more confidence than I'd had mountain biking, I admit, irritated me, and I sometimes forgot that he was new to paddling—like when he dropped onto a wave I was surfing that was plenty big enough for us both, taking up the center and pushing me out one side. With my paddle against his lifejacket I tried to shove him back the other way, yelling "Quit!" His face as I slid off the wave shamed me. Of course he couldn't help it. Dammit, why was it so hard to believe I was the expert out here?

That's happened more than once. Even as he sorted out who I was separate from gender expectations, I learned my own lessons about not pinning certain expectations on Jimmy based on the masculine stereotype of men being more experienced and in control. We often swapped the role of teacher, of excursion planner. Yet I'd still lapse, like on our Alaska trip when I was trying to write and was annoyed at being interrupted by Jimmy's questions. What was I taking in my backpack? How many pairs of socks and water bottles? Did we need to seam seal my tent again since it rained our last time camping? I answered a few before I blurted out, "Aren't you an Eagle Scout?!"

"We stayed in shelters on the Appalachian Trail," Jimmy said contritely. "It's been ages since I used my tent."

"Or a real backpack," I thought to myself, turning back to the desk in the motel room. His mountaineering pack meant I was carrying one-third more than he was, but I hadn't thought to check his gear before we got on the ferry. He'd told me stories about backpacking on the AT. I thought he had experience.

Behind me I could hear his mumbling as he stuffed the tent into the sack. "No, you freaking redneck, you moron, you academic. Don't be so stupid. Just get your non–Daniel Boone-ass out of . . . " That was it. I was busting a gut laughing, and folks walking by the open door were peering inside.

Now don't go getting the idea that Jimmy's perfect. When he visits the Eggers boys back home he still goes riding and drinking with them on the back roads and tosses Miller cans out the window, "'Cause everyone knows you don't want the cops to stop you and there be empty beer cans." And even though over the years I've excelled in my climbing and learned from teachers other than him, I still catch him second-guessing my choices. He's trying to adjust to the fact that I'll lead some climbs he considers too hard or risky, but it's not easy. "You don't want to do that," he occasionally says, which can send me into death spirals of self-doubt. But over the years I am learning to trust my instincts. If I've thought through all the consequences and feel confident, then I let Jimmy know that yes, I *do* want to do that climb or run that rapid. It took him a few years to learn not to prioritize his time outdoors over mine, to not assume he's the only one who defines and refines his selfhood in nature. Though we have our moments, it's stimulating to grow alongside someone who enjoys these activities as much, to compare our cycles of fear and confidence and skill, our insecurities over our weight and strength and body image. To share the work of teaching and writing and thinking about "nature" from a ledge on Donner Summit's School Rock *and* from an academic perch.

Outdoor recreation constructs and commodifies nature to varying extents, and Jimmy wrestles with the moral and practical complexities of our use of these different outdoor spaces. On one hand, we believe in going light and simple. On the other hand, we're suckers for new gear and have a two-car garage that's really a walk-in closet for toys. Jimmy picks up the smallest scrap of trash at the base of climbs and rants about those who cause others to lose access privileges. He argues that wilderness areas aren't the place for bolted sport climbs but should allow fixed anchors. How much are our views shaped by selfishness? Who can ever know for certain? But we test them against each other, discussing and dissecting how we relate to nature through these activities. During the busiest of academic semesters, we still try to set off one day a week for climbing. We call these marriage-maintenance days. In addition to spending time together away from deadlines and dress codes, we maintain our bond by maintaining our bodies—feeling physically good about ourselves enhances our sexuality. But that's not all. Climbing frees the stress from our shoulders and neck that days reading and sitting at the computer bring. It frees our senses to respond to the physical world and releases our detached intellects from being in charge, frees us from other responsibilities so we can revel in our reliance upon each other, and frees us to be animals at the most basic sensual and perceptive levels. And this puts us back in touch with each other and with the larger creation. Then we're ready to go back to teaching students and saving forests.

Jimmy enjoys being a bridge between his buddies who grade roads and his colleagues who grade papers. Even though nowadays he's reading theory in the Wild West, the working-class fella from the South is still there—in the drawl, in the love of a specific place inhabited by his family for generations (there's a Guignard Lane for heaven's sake), in his notion of nature as garden and countryside and home as much as wilderness. And especially in his habit of challenging his students, peers, and professors to consider how much more important the plumber is to society than the intellectual who can't deal with his or her own waste. He goes to motocross out here too, but not as often as he goes to public hearings to protest mines and support roadless initiatives. He writes letters to the editor of *Climbing* magazine urging folks to respect lichen and climb elsewhere. He's morphed into a unique man, one who breaks down binaries such as feminine and masculine, redneck and academic, construction worker and environmentalist, as well as reader and doer.

I like to say that there are two kinds of people in the world: those who think in binaries and those who don't. But the joke is lost on many folks, so the telling of it discourages me. Binaries are reflective of a culture that constructs difference in terms of the logic of hierarchy (Derrida 1981), associating power and dominance with the first term at the expense of the second, thus making oppression seem inevitable or natural (Plumwood 1997). Even Jimmy, as he slays the dragon of dualisms, often views the world through binaries. And as environmentalists tackle local issues we too often divide them into us and them, right and wrong, morally superior liberals and unenlightened conservatives. We will never get anywhere this way because all binaries stem from the separation of humans and nature. As environmentalists and ecofeminists we should not try to reverse which concept is valued more but to challenge all such simplistic separations: humans and animals, men and women, civilization and wilderness. If we don't start recognizing that the world does not automatically lend itself to being dissected into oppositional pairs, that this is a logical structure we have imposed so often it now seems natural, well, there is not much hope for us reuniting with nature or each other. We need to remember that there are almost always more than two options.

Jimmy reminds me of the beautiful complexity of creation as he bounces from one end of the gender spectrum to another, and dallies in the middle. He teaches me about reliance when I want to go on a solo backpack by taking care of the dogs, house, and garden until I return. (When I do go solo, his masculinity is challenged by people who question why he *let* me go alone.) He teaches me how to give up control sometimes—to him, to the weather. Most of all, he reaffirms my belief that there is no perfect man. But breaking down the binaries rather than switching from one extreme to another—from macho ape to 100 percent politically correct new age man—is a better route to reenvisioning the male role in society and in nature. And this is the first step in redefining our roles as husband and wife. Again, we don't swap apron and briefcase and call ourselves enlightened; we break down the binaries until we both can go out and both can stay home and neither feels restricted to one realm or one set of behaviors. By revamping our gender roles outdoors, we learn how to carry those lessons inside. Someday we want to share our love of the outdoors with our kids ("When? When?" ask parents and grandparents over the phone). With luck, when we do number more than two, spending time being active outdoors will help us to redefine our roles as mother and father and to pass on a less binary worldview to our children.

Jimmy has been woven from a history—cultural, familial, and personal—into someone who, to my mind, has the perfect imperfections and many more startling gifts. And these gifts have been most directly revealed to me through our interactions outdoors. It makes me sad to read in climbing books and guides the admonishments not to make your girlfriend your climbing partner. Relationships can't stand the strain, the authors preach. Since I know so many couples out proving that dictum wrong, it makes me wonder why that myth still circulates. Clearly this belief has something to do with how men and women communicate their needs, desires, and fears outdoors. I get the feeling that many guys despair of ever teaching a girlfriend to love an activity the way they do. In their well-meant efforts, these men often impose a certain way of interacting with the outdoors onto the woman they are instructing rather than opening up a space in which she can discover her own way of relating to nature—including her own body. The assumption that there is only one way to love or approach an activity is just as dangerous as the assumption that all men approach such outdoor activities competitively and all women approach them noncompetitively. The natural world can be a space that's liberating for both men and women because they're away from the domestic sphere and associated gender expectations; but for others the confusion over what their roles should be away from the hub of society causes fractures. Or worse, one partner insists on certain gender constructions while another tries to escape them. But is it realistic to think these fractures won't show up eventually in the relationship if they avoid climbing or biking or paddling together? I don't think so. Instead, dealing with such conflicts in the realm of recreation can teach a couple inter-relational skills they can carry back home and unpack.

Sharing physical outdoor activities early in our relationship gave Jimmy and me a crash course in how to express our feelings to each other. This turned out to be extremely important when, after we'd known each other only two months, my dad died. When Mom passed away less than a month later, I knew I wanted Jimmy to come home with me and be there at her service. But, except when climbing, I was not accustomed to asking for assistance, to leaning on someone. I'm convinced it was the trust we'd created through climbing, as well as the good communication patterns, which enabled me to make that request.

"You got me?"

"I've got you."

And though Jimmy was not looking forward to entering such an emotionally charged situation, he took in what slack he could and stood at the base of my grief, holding the rope that connected me to my life and my future, and, when I was ready, lowered me back to his side.

I'm not suggesting that Jimmy is the pattern other husbands or boyfriends of nature lovers should follow, as in, step one: find a southern Eagle Scout; step two: take him climbing. There is no pattern—rather, there are many. I believe the reason Jimmy interacts easily with women outdoors and interacts with nature in a way that's enjoyable for me to be around is that he's aware that both gender and nature are complicated concepts and is interested in exploring how the cultural construction of each affects us—individually and as a couple.

Works Cited

Anderson, Lorraine. Preface to *Sisters of the Earth: Women's Prose and Poetry about Nature*, edited by Lorraine Anderson. New York: Vintage, 1991.

Derrida, Jacques. *Positions*. Translated by Alan Bass. London: Athlone Press, 1981.

Faludi, Susan. *Stiffed: The Betrayal of the American Man*. New York: Perennial, 2000.

Kolodny, Annette. *The Lay of the Land: Metaphor as Experience and History in American Life and Letters*. Chapel Hill: University of North Carolina Press, 1975.

Nash, Roderick. *Wilderness and the American Mind*. 3rd ed. New Haven: Yale University Press, 1982.

Plumwood, Val. *Feminism and the Mastery of Nature*. London: Routledge, 1997.

Shepard, Paul. *Man in the Landscape: A Historic View of the Esthetics of Nature*. New York: Knopf, 1967.

unusual natures

Consuming Cities
Hip-Hop's Urban Wilderness and the Cult of Masculinity

STEPHEN J. MEXAL

Imaging the City/Imagining the City

The recent spate of critical attention paid to environmental concerns has—perhaps unsurprisingly—tended to assume a rather narrow conception of what, exactly, constitutes the environmental.[1] In its popular usage (for example, "trying to save the environment"), the term becomes roughly equivalent to *natural* or is simply used to signify the out-of-doors, the organic. But its etymological roots lie in the Anglo-French *environner* (loosely, to physically encircle), and as such, it is crucial to reconnect our awareness of both the environmental and the ecological to a sociologically based understanding of proximate space.

Proximate space is our immediate, familiar environment: not, for most Americans, a space of trees and streams and hills but rather a space of concrete and glass and steel. For surely this—this prosaic, proximate space of apartments and sidewalks and convenience stores, what Gaston Bachelard calls "really inhabited space"—is a more intimate, personal space than any postcard-perfect mountain or valley (5). According to the most recent available census data, over 80 percent of Americans reside in metropolitan areas, a number that has grown 2.6 percent over the past decade. To that end, in order to better grasp the relationality between the subject and space both distant and proximate, it becomes necessary to expand our conceptions of the environment, of nature, to include things not specifically *natural*.

In his classic work *The Image of the City*, Kevin Lynch emphasizes the reciprocal nature between subject and city. Urban space, he suggests, forms the individual: "[people] extract structure and identity out of the material at hand" (43). Urban space, then, is not strictly an exterior, geophysical concept; it also forms and informs our interior cognitive spaces. As the subject constructs the city, so the city constructs the subject. In this sense, it is useful to think of urban space less as a physical referent than as a series of signifiers: as a discourse, an ideology, a performance. The inner city has become a metaphor, in the popular imagination, for a new wilderness, a cultural development that has significant ramifications for the way in which masculinity is perceived and performed. As such, in this essay I will examine a cultural mode conceived of and executed as both masculine and urban: hip-hop music.[2] In much hip-hop cultural production, there is a deep ambivalence regarding the trope of the urban wilderness, often reflected in the simultaneous acceptance and subversion (in lyrical form) of this trope. This ambivalence not only complicates our understandings of an environmentalist, preservationist approach to space, urban and otherwise, but it also highlights the gendered implications of a geographical and sociological marginalization, making manifest the link between the dispossessed masculine subject and a dislocated urban space.

From Jungles to Ghettos

In most contemporary Green discourse, "wilderness" is something to be saved, to be placed under protection. Andrew Ross, by way of example, notes that "the priorities of mainstream environmentalism are wilderness and natural-area preservation" (15).[3] Historically, though, wilderness has *not* typically functioned as a site of preservation; in the West—excepting, perhaps, in the last half of the twentieth century—the cultural significance of wilderness generally emerged not through the act of conservation but rather through the act of conquest. Viewed historically, wilderness becomes a set of spaces and values oppositionally defined as "not-civilized."

Andrew Light has emphasized what he calls the "cognitive dimension" of wilderness and its role in imagining the "classical view" of not only wilderness but also, by extension, the savage (par. 7).[4] Light argues that, classically, wilderness is constructed by three interrelated theses: (1) It is geographically separate and distinct from nonwild, civilized spaces; (2) its inhabitants are viewed as savage, sub- or non-human beasts; and (3) as a result, civilized humans and their civilizations maintain their separateness and superiority (pars. 11–13). This ideology of wilderness not only discursively constructs and sustains the primacy of civilization over wilderness, it also allows for the annexation of certain politically useful "wild" areas for purposes of "civilizing." To better illustrate the invention of wilderness, I quote at length:

> [T]he wilderness is a place that is always marked as the realm of the savage who is . . . thought to be cognitively, or mentally, distinct from the civilized human. The savage is always marked as the thing that we outside of the classical wilderness, we civilized people, are not. We are superior to the savage simply in not being the savage and

part of what makes us not savage is our possession of reason, our control over our passions. The savage is completely subject to his passions and is in fact driven by them to the point where he may not escape the wilderness. But the escape from the wilderness is not merely a physical escape; if the savage leaves wild nature he is still wild because of the "cognitive" wilderness within him. (par. 7)

That is to say, wilderness is less a marginalized physical place than it is a physicalization of the marginal. And while it may be seen as a cartographic fiction, it is, to be sure, an ideological actuality. It is the very construction of the uncivilized, the unhuman. It is not, Light writes, "the mere physical surroundings but [rather] the claim of those surroundings on the mentalities of its inhabitants" (par. 7). Or, to invoke an old cliché, under an ideology of wilderness, one might take the savage out of the wild, but one can never take the wild out of the savage.

Given the ideological nature of wilderness, coupled with the relative demise of the colonialist aspiration to, in the language of Conrad, bring the "unstained light" (18) of civilization to "the dark places of the earth" (19), it becomes clear that the locus of our contemporary understandings of wilderness (in the classical sense) has shifted. It has moved from the natural to the man-made, from the country to the city. It is the inner city, in specific, that has inherited the metaphorical bequest of classical wilderness: it has constructed a new set of boundaries; it has invented an urban wild, a wilderness of the ghetto.

This wilderness, however, is not a wholly imaginary one; it exists as a series of tropes projected from and onto the real circumstances of urban existence. As such, it is necessary to draw a distinction, however blurry, between urban material conditions and urban discourse. The former is concrete and geographic, the latter, a cognitive abstraction. Yet the two are mutually implicated; urban policy can and does spring from the wells of urban ideology, and by schematizing the inner city as "wild" or "not-civilized," policymakers and urban planners of all political persuasions have created a series of inner city strategies rooted in historical conceptions of wilderness, emphasizing the importance of conquest, conservation, or rescue/recovery. Under this logic, "urban renewal" or gentrification serves to conquer or reappropriate the untamed urban space; welfare, depending on its political proponent or opponent, functions as either an attempt to conserve the urban status quo or to rescue the denizens of the inner city. Given the association between urban ideology and urban policy, it is critical, in our understandings of urban-based culture, to work to distinguish the urban performed from the urban real. Accordingly, while many critics use the terms "inner city" and "ghetto" interchangeably, I suggest a slightly different usage. I use the phrase "inner city" to denote the actual physical geography, while "ghetto"—because of both its negative connotations in mainstream discourse and its reappropriation by hip-hop artists—I use to signify the ideology of urbanity and the social performance arising out of the geography of the "wild" inner city.

Where the Wild Things Are

The urban wilderness is, to even casual observers of twentieth-century culture, certainly not a new concept. The most immediate example, perhaps, is Upton Sinclair's *The Jungle*, with its equation of urban and proletarian wretchedness with the classical conception of wilderness implied by its title. It is a trope that is also present in texts such as John Howard Griffin's *Black Like Me* ("hell," Griffin's narrator writes of the inner city in which he finds himself, "could be no more . . . agonizingly estranged from the world of order and harmony" [67]) and Richard Wright's *Native Son*, as well as films such as Ridley Scott's *Blade Runner* or John Singleton's *Boyz in the Hood*.

But to theorize the urban wilderness is to theorize a paradox. It is an amalgamation of the settled and the untamed, straddling the cognitive gap between a pure, imagined wilderness—the free space of an endless desert or forest—and the gridded, heavily restricted space of a modern urban geography, the space of calculated city blocks, well-timed stoplights, and extensive zoning laws. As such, the trope of the urban wilderness serves as a metaphoric bridge between free (wild) and restricted (city) spaces; it is, of course, the very essence of urbanity, but it also signifies the modern wild imaginary. Due, then, to the conflicting discourses of a smooth, unregulated wilderness and a rigid, systemized urbanity, the urban wild becomes—like Rudyard Kipling's jungle or Zane Grey's wild West—a fraught, dangerous space. The inner city, in part because of its large numbers of poor and minority residents, has a complicated, delicate relationship with its whiter, wealthier urban/suburban counterpart, the city-at-large. For the inner-city denizen, the city-at-large is the object of geographic affinity as well as social antagonism: racial and economic conditions can change drastically by moving only a few blocks. The inner city thus becomes a symbolic space charged with racial and economic tensions, a space outside space. As such, the ideology of wilderness dictates that it be fenced off, that it be restrained as a site of geographic and political domination.

Because of this restraint and domination imposed on the space itself, inner-city identity can become a process of negotiation, of performance, whereby social roles are mediated and constructed through the discourse of urban space. That is to say, because the inner city is geographically separated and troped as "wild," identity formation is often contingent on an understanding of the social and cultural marginalization of that same "wilderness." The inner city, then, is a marginalized space; it is not the geographically and economically centralized space of commerce and retail but rather the peripheral locus of a set of values oppositionally defined as other: not-white, not-wealthy, not-educated. This social, economic, and geographic marginalization, enabled by the trope of the urban wilderness, gets translated into a social performance that Chuck D, a rapper from the group Public Enemy, calls a fantasy. "As black people," he notes, "we ain't in control of our reality. When you're not in control of your reality, fantasy becomes a bigger influence" (McCall 48). This willingness to embrace the performative aspects of identity, to be influenced by "fantasy," has significant implications for the way in which masculinity is constructed within the discourse of the inner city. Hip-hop culture works to negotiate between the performative aspects of the urban wilderness and the social realities of the inner city to establish, ultimately, a defense against inequitable material conditions and the discourse of the urban wilderness.

Performing the Ghetto

Hip-hop originated in the boroughs of New York in the mid-1970s, a culture birthed in tiny clubs and city parks. Musically, it is perhaps best understood as pastiche; the rhythmic/musical backing tracks were composed of small fragments of preexisting recordings, manipulated first manually, with two turntables and a mixer, and later digitally, in a computer or sampler.[5] The prevailing compositional technique in the 1970s and 1980s was to find the rhythmic core (or "breaks") of two or three recognizable records isolated of their lyrical content and to manipulate portions of those decontextualized sounds into a new aural landscape that would then, through rhythm and repetition, create a new musical point of convergence for the singular black urban voice. Tricia Rose, in her book *Black Noise*, elaborates on this process of what would come to be identified as sampling:

> Sampling in rap is a process of cultural literacy and intertextual reference. Sampled guitar and bass lines from soul and funk precursors are often recognizable or have familiar resonances.... [T]hese samples are highlighted, functioning as a challenge to know these sounds, to make connections between the lyrical and musical texts. It affirms black musical history and locates these "past" sounds in the "present." (89)

The elements of rhythmic/musical familiarity and repetition, coupled with the complete lack of vocal melody, thus serve to accent the significance of the performing subject. And the performance of the subject, in turn, highlights three principal themes: (1) the black, (2) the masculine, and (3) the urban. Hip-hop, like much historically androcentric African American art and literature, tends to conflate the tropes of blackness with the tropes of masculinity.

One of the first studio recordings of hip-hop music, "Rapper's Delight," a fifteen-minute unstructured opus released in 1979, goes to great lengths to emphasize the geographic positionality that produced its members. The song was recorded by the Sugarhill Gang and released on Sugar Hill Records, two rather unambiguous references to the Sugar Hill district of Harlem in upper Manhattan. This allusion to geography serves at least two purposes. First, it emphasizes the significance of place in urban discourse; in this sense, one's socioeconomic location becomes conflated with, if not subsumed by, one's geographic location. Second, it works to reimagine that same urban geography, which is to say that the Sugarhill Gang is not, obviously, a neighborhood itself but rather a cultural performance of a neighborhood; it is a discursive re-presentation of an urban actuality. In this sense, it is an ideological reconstruction, encapsulating the Althusserian notion of ideology as the representation of "the imaginary relationship of individuals to their real conditions of existence" (67). Since its inception, then, hip-hop has always embraced an urban ideology, a manifestation of the urban real through the lyrical imaginary, the material city through the performed city.

This fusion of the urban real and the lyrical imaginary arguably reached its zenith in the late 1980s, a time when hip-hop began to formulate a lyrical counternarrative in response to the Reagan-Bush neoconservative culture-of-poverty hegemony. It was a time that also, not coincidentally, saw the rise of "gangsta rap" and the emergence of the theater-of-terror hip-hop persona: the hypermasculine, hyperviolent, hypersexual embodiment of white America's latent racial anxieties: Bigger Thomas with a beat.

Perhaps the most salient example of this phenomenon is N. W. A.'s seminal 1988 album *Straight Outta Compton*, a work that all but invented the so-called "gangsta rap" subgenre. The record's significance—not to mention its popularity—was no doubt due in part to the fact that the five-man group was actually from the inner city and rapped about urban existence with a rage and a (presumed) veracity unseen theretofore. The record begins ominously, with a menacing voice, devoid of any backing music, intoning, "You are now about to witness the strength of street knowledge." Before the album proper has even begun, then, the significance of an urban ideology is suggested. "Street knowledge" stands here in implied contrast to "school knowledge," such that the cultural ideology of all that is outside the inner city—the city-at-large, the suburban—is dismissed at the outset. In its place, we are introduced to three narrators—Ice Cube, Eazy E, and MC Ren—who seem to delight in embracing the discourse of the "urban savage." In their cultural construction, inner-city existence produces a type of insanity, a cognitive disengagement that is, they imply, a direct result of residing in Compton, which is a notoriously poor area of Los Angeles but hardly the most noxious example of the American inner city. Regardless, references to mental instability from the title song, "Straight Outta Compton," include such phrases as "crazy-ass nigger," "crazy as fuck," and a "crazy motherfucker from the street." Here, the specifics of urban place—"straight outta Compton"—effect a severe sort of cognitive alienation that seems to directly conform to the subject's inability to map himself in a geographic sphere larger than his proximate urban environment. Even a material as simple as asphalt ("the street") is central in the metonymic construction of personal geography, both psychic and physical ("a crazy motherfucker from the street"). Significantly, despite at least forty-nine references to "Compton" on the album *Straight Outta Compton* as a whole, there is virtually no mention of the geographic particulars of Compton itself. That is to say, despite fleeting references to "1st and 15th" ("Dopeman"), "Crenshaw [Boulevard]," and the "East side, south of Compton" ("8 Ball [Remix]"), the geographic specifics of the inner city are ultimately downplayed in favor of the cultural specifics of the ghetto. "Compton," then, becomes a floating signifier, made to stand for any inner city, anywhere.

Violence, in addition to mental instability, plays a significant role in the group's engagement with the trope of the urban wild. For theirs is not, it would seem, the latent, implied violence of social protest but rather the explicit, naturalized violence of the wild. Violence, here, is less subject than object, less a signified than a signifier. "Straight Outta Compton" is, Ice Cube's narrator asserts, a "murder rap," but this is not really the case. It is not *about* murder; violence, instead, acts as yet another sign, much like insanity, that ultimately defers its signification to the referent of inner-city geography. The text of the same rap includes violent references to "bodies [being] hauled off," "a crime record like Charles Manson," and the use of an AK-47 as a "tool." Here, violent imagery is largely incidental to their construction of the specifics of a discourse of the ghetto. That is to say, the text of the rap is not necessarily about violence per se but rather about the geocultural construction of the inner city. Each verse begins with its narrator shouting the title and refrain ("straight outta

Compton"), and the chorus is comprised of the eerie sampled chant "city of Compton, city of Compton." Inner-city geography, then, is accentuated, but an inner-city discourse is of preeminent import.

However, the very energy by which this theme—of the violent, mentally unstable urban savage unfit for mainstream existence—is pursued seems to undercut its impact. What matters here is less the psychic damage of the modern inner city but rather the construction of that psychic damage. The inner city thus becomes a performance; we become witness to a "theater of the ghetto," wherein urban space is experienced but also performed. Geography, as such, is reconfigured as discourse.

This discourse is an emphatically masculine one, but it is a unique masculinity, a masculinity that is inseparable from the trope of the urban wilderness. Here, the male figure assumes the role of the savage, the predator, one who is questionably sane and lives by a code of random violence. This figure is also, not surprisingly, one that is deeply misogynistic. The interrelated tropes of blackness and masculinity, Michelle Wallace writes, allow "for only the most primitive notion of women—women as possessions, women as the spoils of war, leaving black women with no resale value" (676). Much like the motifs of violence and insanity, the trope of the "black macho" in "Straight Outta Compton" is significant for the zeal by which it is embraced. Examples include references to a "good piece of pussy," a "dirty-ass ho," and "the bitch who got shot." The conflation here of "bitch" and "woman," coupled with the prevalence of the word "motherfucker" throughout, suggests an easy espousal of the tropes of the black macho. While I would not seek to excuse or justify the group's misogynistic tendencies, it is important to contextualize them, to place the themes of violence, insanity, and misogyny within the larger discursive framework of the urban wilderness.

Which is to say, my argument is one of tropology, not one of anthropology. It would be erroneous to use the text of this or any other rap as a sort of simplified urban character study, an inner-city anthropological unearthing. To do so would ignore the very real elements of performance and linguistic play at work here. When Eazy-E raps that he is "a brother that'll smother your mother / and make your sister think I love her" ("Straight Outta Compton"), the violence primarily serves to update and recontextualize, in lyrical form, the African American tradition of the dozens, or the humorous insult. More to the point, the word "smother" is present largely because it rhymes with "mother." This sharp injection of a specific type of African American humor—the "your mother" joke—against a backdrop of misogyny, violence, and insanity serves to undercut and, I think, resituate those same elements. It suggests a self-awareness (surely, the reflexive nature of humor implies that the object of the joke is Eazy-E himself as much as "your mother"), and that self-awareness, in turn, carries over to their themes of misogyny, violence, and insanity. Given this, these themes, because of their link to classical wilderness, reflect a subversive engagement with the trope of the urban wilderness and its creation of the urban savage. It is a *conscious* application of the trope, employed in order to reveal its constructed nature. And by virtue of the fervor by which that trope is embraced—the motifs of violence, insanity, misogyny, and profanity are inarguably spectacularized; the word "fuck," by way of example, appears an astounding 134 times on the sixty-minute album—its ideological foundations are exposed.

This is not, however, to imply that this social performance of the ghetto is largely or even partly divorced from the material conditions of the inner city. Instead, it is *through* the discursive, spectacularized performance of urban space that hip-hop works to illuminate the real-world material conditions of the inner city. "The 'ghetto,'" writes Robin D. G. Kelley, "continues to be viewed as the Achilles' heel in American society, the repository of bad values and economic failure.... [The] cultural and ideological constructions of ghetto life have irrevocably shaped public policy, scholarship, and social movements" (9). As such, the act of "performing the ghetto" has significant implications for our understandings of both urban wilderness and urban masculinity. Much like the Fool, the figure of the rapper functions as the knowing outsider, exposing ideological fallacies through his assumed status as social outcast and spectacle.

Boyz in the Hood

In the past ten years, hip-hop culture has taken a turn away from the politics of the inner city; in place of a strategic performance of the ghetto, one now often finds 1980s-style materialism. No longer are poverty, joblessness, and social inequity the topics of popular rap music. In their place are accounts of luxury automobiles, Rolex wristwatches, and expensive champagne. The politics of the inner city, it would seem, have been supplanted by the politics of consumption. However, there are a number of exceptions to this trend; at least one of them is the late Tupac Shakur, who, under the name 2Pac, worked to update and even append a mythological dimension to the trope of the urban wilderness. In "Outlaw," his narrator's attitude toward both the inner city and the urban wilderness vacillates between desire and disgust. "I broke the law," he raps, "and ... before I close my eyes, I fantasize I'm living well / then I awake and realize I'm just a prisoner in hell / . . . / Will I last? Heaven or hell? Freedom or jail?" Because of its marginalized status, the urban wilderness provides, as suggested by the fantasy of "living well," the opportunity for illicit material gain but ultimately only allows for the socioeconomic and geographic position of a "prisoner." Throughout, his defiantly masculine and self-consciously savage narrator rails against "bitches" and "hos" (whores), and as his discursive ghetto becomes littered with "dead bodies," "gunfire," and "homicide," violence and masculinity gradually become subsumed within the larger tension between the heaven/hell, freedom/jail binaries. Over the course of the song, then, masculine/feminine, violence/peace, urban/nonurban, and heaven/hell eventually become equivalent oppositions. Here, the urban wilderness is both a cognitive abstraction (heaven/hell) and a tangible, material place (freedom/jail). Along with other recent rappers such as DMX, 2Pac bridges the disconnect between discourse and praxis, emphasizing the material reality belying the trope of the urban wilderness as well as the interdependency between the ideology of wilderness and the spectacle of masculine violence.

In her discussion of the sexual politics of hip-hop culture and rap music, Tricia Rose notes that "many black male rappers expose the vulnerability of heterosexual male desire in their exaggerated stories of total domination over women" (172). This I would amend only slightly: male rappers disclose the vulnerability of masculinity, period, in their hyperbolic narratives of domination over women and urban space. The ambivalence—the simultaneous acceptance and subversion—of the trope of the urban wilderness not only creates a stage for the social performance of a certain theater of masculinity, it also constructs the theater itself. The discourse of the inner city both creates and is created, consumes and is consumed. It is invented through the subject-ive enactment and espousal of the trope of the urban wilderness, but paradoxically, it also invents the very possibility of that subject. In this sense, the figure of the rapper exists in an interdependent, reciprocal relationship with the urban wilderness. This trope, coupled with the performative implications of the inner city, invents a space for a dispossessed hyperbolic portrait of the masculine, constructing, ultimately, a cult of masculinity.

This "cult," though, subverts our usual understandings of the term as an extremist or separatist religious sect. The word derives from the Latin *cultus*, for "worship," by way of the past participle of *colere*, the verb "to cultivate." And it is in this sense—of cultivation—that I intend the term. It is a performance of masculinity that is cultivated for the strategic purposes of exposing the racist ideology that belies the trope of the urban wilderness. The neologism "wilding," with its roots in hip-hop culture, might help illuminate this idea of a cultivated masculinity.[6] Loosely, "wilding" might be defined as the deliberate creation of social disorder, although the *Oxford English Dictionary* in a 1993 supplement defines it as the "action or practice by a gang of youths of going on a protracted and violent rampage in a street, park, or other public place, attacking or mugging people at random along the way." The term, Robin D. G. Kelley argues, "reveal[s] a discourse of black male youth out of control, rampaging teenagers free of the disciplinary strictures of school, work, and prison" (53). Similarly, Houston A. Baker notes that it sounds "ominously uncivil and antipublic" and views acts of "wilding" as the manifestation of an authentically or foundationally wild youth that stands in contrast to the "culturally constructed" wilderness of the park (46). Baker traces the term to the Central Park jogger rape case of 1989,[7] an argument that becomes all the more salient in light of the recent exculpation of the five African American and Hispanic young men convicted of the crime. In a confession, the five characterized their alleged participation in the horrific rape and near-fatal beating as "going wilding," an application of a trope that gives their exoneration all the more significance. "Wilding," in this sense, stands as the preeminent exemplar of the urban wilderness, as the creation of social disorder under the discursive auspices of embracing the trope of the urban wilderness. It is, however, a trope that is embraced strategically, in order to subvert its resonance. "Wilding," as applied to the rape case, is suggestive of the *idea* of social disorder rather than its actual practice. That is to say, the coerced confessions reflect an acceptance, on the part of the investigators, of the trope of the urban wilderness. Presumably, the explanation that sounded most accurate was that the young urban residents were, in fact, "wild" and had accordingly gone "wilding," wreaking their savagery on Central Park. In hindsight, "wilding" is suggestive not of an actual crime but of an ideological maneuver, a performance of the ghetto that reveals the racist ideology buttressing the trope of the urban wild. To be sure, there was no act of wilding that culminated in a rape that night. The five alleged rapists (now acquitted) employed the tropes of the urban wilderness in their (false) admission of guilt, which retrospectively exemplifies the way in which masculinity and urbanity are troped in contemporary discourse. "Wilding" is thus an ideological signifier; to employ it both sustains and subverts the tropes of the urban wilderness and the myth of the hypermasculine inner-city savage. This has potentially significant implications for the way in which we schematize and perceive all types of wilderness, both natural and urban. The ideological dimension of space may perhaps be inescapable, but it should nevertheless be carefully considered. Some types of wilderness, it would seem, are not worth conserving.

Ultimately, the ambivalence regarding the trope of the urban wilderness seen in much hip-hop culture is suggestive of a complex, interdependent relationship between the dispossessed masculine subject and the cognitive and geographic space of the inner city. Hip-hop music enacts a social performance of the ghetto that, in its lyric reimagination of the inner city, contrives a deliberately exaggerated cult of masculinity. This cult, though, is always/already constructed in a tenuous reciprocal relationship with the trope of the urban wilderness; in this sense, masculinity both consumes and is consumed by the ideology of the inner city. As such, this network of ideological signifiers constructs, through hip-hop culture, a performative space for a disenfranchised cult of masculinity, established as a discursive defense against oppressive material conditions both socioeconomic and geographic and enacted against the cognitive and physical backdrop of the urban wilderness.

Notes

1. Rosedale's *The Greening of Literary Scholarship* well illustrates the growing prevalence of an ecocritical awareness among literary theorists and Left intellectuals; along the same lines, membership in the Association for the Study of Literature and the Environment (ASLE) has increased substantially in the last several years.

2. Generally speaking, "hip-hop" is defined as the musical and cultural aesthetic of rap music, whereas "rap" is the spoken vocal performance in specific.

3. For more on the use value and ideology of wilderness, see John Ray Knott's *Imagining Wild America* or Callicott and Nelson's *The Great New Wilderness Debate.*

4. This notion of a "cognitive dimension" of wilderness is a phenomenological concept that seems closely aligned with Frederic Jameson's theories about postmodern hyperspace and the impossibility of the subject cognitively "mapping" itself into the contemporary urban landscape. For more on this, see Jameson's *Postmodernism,* 44–45, 54.

5. My argument here is an odd paradox: the historically specific generalization. While I acknowledge that (a) hip-hop musical production has obviously changed over the decades, and (b) there was and is no single production technique, in the interest of brevity, I do not address these issues.

6. At least one early use of the term in recorded rap music is found on a 1988 Ice T album, *Power.* In "Radio Suckers," he raps about "gangs illin', and wildin', and killin'," ultimately, in the narrative, situating the act of "wildin'" as a strategic response to inequitable socioeconomic conditions.

7. This is actually incorrect; the term had been in use as hip-hop slang for at least a year or two prior to the rape. For more, see Robert F. Worth's "A Crime Revisited." For a contemporary overview of the 1989 crime and the 2002 exculpation of the accused, see McFadden and Saulny.

Works Cited

2Pac. "Outlaw." *Me against the World.* Interscope/Jive, 0841-416362. ℗ 1995.
Althusser, Louis. "Ideology and Ideological State Apparatuses." In *Lenin and Philosophy.* New York: Monthly Review Press, 1971.
Bachelard, Gaston. *The Poetics of Space.* Translated by Maria Jolas. Boston:

Beacon, 1994.
Baker, Houston A., Jr. *Black Studies, Rap, and the Academy.* Chicago: University of Chicago Press, 1993.
Callicott, J. Baird, and Michael P. Nelson, eds. *The Great New Wilderness Debate.* Athens: University of Georgia Press, 1998.
Conrad, Joseph. *Heart of Darkness.* Boston: Bedford/St. Martin's, 1996.
Griffin, John Howard. *Black Like Me.* New York: Signet, 1976.
Jameson, Fredric. *Postmodernism: Or, the Cultural Logic of Late Capitalism.* Durham: Duke University Press, 1991.
Kelley, Robin D. G. *Yo' Mama's Disfunktional!: Fighting the Culture Wars in Urban America.* Boston: Beacon, 1997.
Knott, John Ray. *Imagining Wild America.* Ann Arbor: University of Michigan Press, 2002.
Light, Andrew. "From Classical to Urban Wilderness." *Trumpeter* 9, no. 4 (1992): 23 pars. http://trumpeter.athabascau.ca/content/v9.4/light.html (accessed 7 June 2003).
Lynch, Kevin. *The Image of the City.* Cambridge: M. I. T. Press, 1960.
McCall, Nathan. *What's Going On: Personal Essays.* New York: Random House, 1997.
McFadden, Robert D., and Susan Saulny. "A Crime Revisited: The Decision; 13 Years Later, Official Reversal in Jogger Attack." *New York Times,* 6 December 2002.
N. W. A. "8 Ball (Remix)." *Straight Outta Compton.* Priority/Ruthless, CDL57102. ℗ 1988.
———. "Dopeman." *Straight Outta Compton.* Priority/Ruthless, CDL57102. ℗ 1988.
———. "Straight Outta Compton." *Straight Outta Compton.* Priority/Ruthless, CDL57102. ℗ 1988.
Rose, Tricia. *Black Noise: Rap Music and Black Culture in Contemporary America.* Middletown: Wesleyan University Press, 1994.
Rosedale, Steven, ed. *The Greening of Literary Scholarship: Literature, Theory, and the Environment.* Iowa City: University of Iowa Press, 2002.
Ross, Andrew. Interview. "The Social Claim on Urban Ecology." In *The Nature of Cities,* edited by Michael Bennett and David W. Teague. Tucson: University of Arizona Press, 1999.
Sugarhill Gang. "Rapper's Delight." *The Sugar Hill Records Story.* Rhino/Sugar Hill Records, R272449. ℗ 1997.
U.S. Census Bureau. "Percent Change in Metropolitan and Nonmetropolitan Populations, by Region and Division: 1990 to 1999." http://eire.census.gov/popest/archives/metro/ma99-map1.gif (accessed 10 December 2002).
Wallace, Michelle. "Black Macho and the Myth of the Superwoman." In *The African-American Archive,* edited by Kai Wright. New York: Black Dog and Leventhal, 2001.
Worth, Robert F. "A Crime Revisited: Wilding; A Word That Seared a City's Imagination." *New York Times,* 6 December 2002.

WILD TIME
Prisoners and Nature

KEN LAMBERTON

From my upper bunk at the Tucson prison complex, a narrow window allowed me a southern exposure of the desert outside my cell. An expanse of razed ground, marked with an artificial horizon of galvanized steel webbing, filled the lower two-thirds of the frame. But beyond the fence, an entire basin of creosote, mesquite, and cholla leaned up against the hunched shoulders of the Santa Rita Mountains at the Mexican border. On some evenings, coyotes would call to me with borderless voices from the desert's fringe, complaining about their vagrant allotment in life. I would have gladly traded them places.

Nature writer and poet Alison Deming says we need to outgrow the childish notion that nature takes place only in wilderness. Ecologists understand that cities, because of their potential for varied habitat and abundant food and water, attract wildlife. Tucson, a metropolitan sprawl of more than half a million people, draws coyotes to its urban parks and weed lots. And more: coyotes and deer and javelina, mountain lions and even an occasional black bear. Over the last decade, researchers at the University of Arizona have been studying a city influx of raptors, birds of prey including great horned owls, red-tailed hawks, Cooper's and Harris hawks. These are not only occasional visitors. They are staying, nesting in the eucalyptus, pine, and palm trees that stretch shadows across residential streets and neighborhood parks. The predators feed on songbirds and rodents lured to backyard feeders and birdbaths. They raise their broods above picnicking families and little league ball games. It's happening in Tucson, and it's happening in other concrete and steel wildernesses like Seattle, Chicago, and New York: Nature.

Wendell Berry says that the whole world is wild, and "all the creatures are home-makers within it, practicing domesticity: mating, raising young, seeking food and comfort." Anne Matthews, in her book *Wild Nights: Nature Returns to the City*, writes that "Wild does not always mean natural; urban is not the same as tame. Even in Manhattan, you are never more than three feet from a spider."

In prison, the spiders share your pillow.

Until my release in September 2000, my wilderness was a limited geographical area, not one bound by mountains or rivers or oceans but by chain link and razor wire. My wilderness was a prison. All the same, it was a wilderness with its own nuances of seasonal change, summer droughts and winter freezes, rain, dust, and wind; with its own microcosm of weeds, trees, birds, and insects. Nature was there as much as it is in any national park or forest or monument, those places we always list whenever we speak of wilderness.

Few people would consider that nature can be found in prison. They think that prisoners wake up every morning as bodies on mattresses that move through pointless days, bodies at work raking rocks, bodies at meals, bodies in front of TVs, bodies that undoubtedly live without participating in life. This may be true, but only for those few who lack emotion. There are many locked up who see beyond the concrete walls and scraped earth—or see into it, between the cracks—those who notice wilderness, feel its moods, hear its migrations, sense its shiftings and pulses. Those who actually *feel* a sense of place . . . in that place.

The capacity to connect with one's feelings seems fundamental to embracing nature in a place where nature is drawn out into seams and cracks. If it's true, as some say, that the human soul comprehends no boundary, no edge, then emotion must be the nerve-tangled pathway that wires us to wildness, and the current must flow in both directions. Nature accesses the souls of the hardest criminals, finding weaknesses and breaching barriers, building nests and rearing offspring. Place windows facing toward the migrations of swallows, and men will count the birds. Open cell doors to the trespasses of toads and tarantulas, and men will adopt them. Plant trees, and men will sit under them. Even in the deepest prison holes where we keep society's worst offenders, men will attune themselves to the cockroaches.

I spent nine years of my twelve-year sentence inside the chain-link fences at the Santa Rita Unit in Tucson, Arizona. The prison housed more than nine hundred men inside four "yards"—cell blocks of ninety-six two-man cells—and one "tent city," a group of a dozen ten-man, canvas, military leftovers. A half-mile dirt exercise track circled a soccer field and what we called "the park"—a pocket oasis of evergreen peppertrees, rosebushes, desert willow, and flower beds. Everywhere, even outside our cells on the separate yards, the trees and grass and blooms not only masked the stench of fear and desperation but cut holes in the fences and broke open the stark, gray walls, softening the heavy aspect of the prison.

Now the trees are gone and the men remain locked in their cells. But before Santa Rita went to twenty-four-hour lockdown status, men walked the exercise track in soaking monsoon storms or rested on sun-glazed grass among roses and mint. They hand-fed celery scraps to cottontail rabbits, tossed bread crumbs to grackles and blackbirds. They sat at picnic tables with doves perched on their shoulders. The men knew the intrinsic value of nature lying against the skin—even if they were unaware of how deeply nature affected them.

Tony was a former cellmate of mine and friend for many years who, at the age of sixteen, killed two people in a fit of uncontrollable rage. Arizona prosecutors spared him the death penalty because of his age and sentenced him, instead, to twenty years. He served every day of it, finally earning his release only a few years ago. Last I'd heard, Tony lives a quiet life in a Phoenix suburb.

For Tony, it was the toads that connected him to the place where he experienced a different kind of human emotion, not outbursts of anger or hatred—the only acceptable emotional expressions in prison—but the transformative feelings of regret, even remorse, what many inmates considered signs of weakness. Toads made Tony break weak.

I remember one morning when he returned to our cell from the shower with a cleaned-out peanut butter jar held in one thick, barbell-callused hand. Inside the jar squatted an avocado-skinned Great Plains toad with golden eyes. "It was in the stall," he explained. "You should take it to the park and let it go before someone kills it."

Tony wouldn't touch the amphibian, and I might have dismissed his apparent compassion for the animal had not a second incident with another toad happened later that evening. It was early spring, and the inmate landscape crews had begun to regularly flood the shrubs and annuals growing in three tiers of concrete planters outside our cells. In the evenings, the pooled water would call the toads to the surface to trill for mates or to sit on the runs and wait for beetles to stumble within tongue-reach. Usually the toads were safe. Occasionally, men would amuse themselves by feeding various tidbits to the giant amphibians. The toads would lap up anything tossed to them: insects, smaller toads, popcorn, even smoldering cigarette butts—a particular cruelty that the toads immediately rejected, pulling in their eyes, backing away, and ejecting the obscenity with their tiny hands.

Inside our yard, however, someone was doing more than offering the toads a smoke. Someone was throwing the toads into the coils of razor wire strung along the walls high above our cells and leaving them there to hang and die in the sun.

"This one's too big for the jar," Tony told me that evening. "You better get it quick." At the end of our run, a Sonoran desert toad the size of a ripe mango had pressed itself into a corner of the steps. It was a two-handed toad. "The park isn't safe. Take it and put it under the fence."

At the time, I had only celled with Tony for a few months, but I began to see a different side of him. I didn't know about his crime, only that it was violent and that he'd come to adult prison from a unit for minors after his eighteenth birthday. I knew that he was an A student with the rare kind of family who continued to visit him, even after two decades of confinement. And I knew he had issues with anger, and that he occasionally wrote poetry to express these feelings. "I did a horrible thing," he told me one night as we lay in our bunks after lights-out. "And the kids in Minor's admired me for it."

Tony never came closer to exposing his remorse to me—he didn't have to. I saw it on his face the day he carried that first toad into our cell. Tony was anything but a nature lover. He complained about the mosquitoes. He thought someone should poison the mice that worked every concrete seam for crumbs. He hardly noticed the swallows that nested every year at the end of our run. But the toads reminded him of a past life, of childhood summers with pails of tadpoles following a monsoon rainstorm, of oblate, goblin-eyed amphibians tucked into coverall pockets. The toads in our cell tied Tony to a childhood before he was sixteen, and to a later childhood that never was but might have been.

Years before I celled with Tony, I became friends with a young gay man who had a beautiful tenor voice. On Saturday evenings in the prison chapel, I played guitar and together we led our small church group, David harmonizing with me on songs like "Amazing Grace," "This Is the Day," and "There's Power in the Blood." We also worked together, he as an education clerk, I as a teacher's aide.

When I knew him, David was halfway through his eighteen-year sentence. The men called him "Spot" because of a large burgundy birthmark on the side of his face and neck, and David seemed to enjoy the attention his nickname gained him. I never used his nickname, but there were many who did, including David himself.

In the spring, as temperatures rose and the dark earth in the inmate park relaxed enough to allow dormant bulbs to push up pale green stalks and flower buds, and the perennials began cloaking themselves with blooms, I would find David perusing the baby's breath, yellow columbine, poppies, carnations, lilacs, and irises. Later, when the pipevine swallowtails and queen butterflies arrived to garnish the blooms, David would enlist my help to capture these delicate insects with negligee wings. We used our shirts, throwing them like fishermen casting nets in shallow water. He placed his specimens between sheets of newsprint and dried them in his office under several cases of photocopying paper, the weight pressing the flowers and insects into thin films of color, which he then arranged between twin sheets of adhesive vinyl and trimmed to postcard size.

As I think about it now, I wonder if David wasn't appreciating nature so much as exploiting it—mailing plastic-entombed flowers and butterflies to his family and friends to regain their favor. I like to think that he did it to express the sorrow and shame he felt inside but couldn't find words for. And that I aided him in some small way. Maybe. I can only guess at what made him preserve those small moments of beauty in that ugly place and share them with the people he both loved and hurt the most.

I haven't seen David for eight years, not since the Arizona Department of Corrections shipped him to another prison unit. My daughters still have a few of his plastic flower arrangements, ones he gave me to send home to them.

Steve came to Santa Rita in the early 1980s when the unit first opened. He was thirty years old, and he'd just spent the previous few years in "the Walls," a concrete enclosure of tiered cellblocks at Florence, Arizona, and one of the toughest places in the state to do time. There, prisoners rarely saw the sky. Coming to Santa Rita from Florence was like getting released.

When I met Steve after my conviction in 1987, he had a garden in a corner of the yard near his cell. On warm weekends when he wasn't working at the sign shop, he would drag a hose to his plot of strawberries, bell peppers, jalepeños, and tomatoes, strip off his shirt, and drop to his knees. Officially, Steve's garden was a policy violation, but the guards generally ignored the infraction. The inmates understood its produce belonged only to him.

As we became close friends, I often helped Steve obtain seedlings from the greenhouse, a framed-glass structure at the corner of Yard One in which the now-defunct horticulture program grew trees and shrubs and flowers for landscaping the prison complex. Together, we planted umbrella plants and a three-foot chinaberry tree to shade his vegetables. Even now—two years after my release and more than a decade past those first years—whenever I recall that garden the memories still come with feelings of serenity. But for Steve, serenity and anxiety fought for space in his mind. He never knew his father. He had no family, only a girlfriend who wrote him letters and visited occasionally. His mother drove her car off a mountain highway soon after Steve went to prison, leaving behind a note explaining her intentions. Steve always blamed himself. Why wouldn't he? As the anniversary of his mother's suicide approached every fall, Steve withdrew from his friends and spent more time in his garden, his only place of refuge besides his cell.

The Department transferred Steve back to Florence after the policy of tree-cutting came to Santa Rita, and the guards converted his garden to pea gravel. I can't ignore the coincidence. A few of us managed to salvage the chinaberry tree, transplanting it into the inmate park where it survived for a while longer with a few peppertrees. Whenever I walked near it, I remembered Steve; his chinaberry tree was a relic of better times, for Santa Rita and for my friend.

Although I had trained as a biologist, prison narrowed my education to a more intimate study of nature. The hostile landscape warned me to be unobtrusive and quiet, yet alert. To notice things. Like the ants: those single-sexed, look-alike workers that crawled about the prison grounds in hordes, carrying out menial jobs as inmate masons and carpenters and waste-disposal crews, inmate farmers and food-handlers. They obeyed without thinking, doing what they were told, sustaining a colony that existed to devour budgets and breed an ever-larger workforce of mindless slaves. I couldn't escape the parallels. And the raptors: like the owls that mysteriously violated the perimeter fences to kill and feed on the mice, and to restore my faith in things wild, supernatural. And the trees . . .

On every lap during my daily exercise walks, I passed the length of the inmate park, and, although my thoughts may have distracted me elsewhere, my eyes always turned to the trees. They, like other bits of wildness confined or visiting there, expounded some vital part of me. In the same way that they broke down the walls and cut holes in the fences, those trees lessened the burden of that place.

One tree in particular held special significance for me: a mesquite tree that I had watched over the years grow from a transplanted stick into a humid, bug-clicking canopy of wrought-iron branches and hard green leaves despite successive seasons of mindless pruning. I used to sit against the mesquite's sun- and wind-furrowed trunk and breathe in the smell of its obstinacy. The mesquite was tough, a survivor. It had overcome years of butchery, remaining surprisingly robust. Most springs, it produced a crop of seedpods, which drooped in clusters like blonde dreadlocks; some winters it even held onto its leaves. But the mesquite grew askew, hunching away from a nearby building as if crippled by the heaviness of its gray walls.

The mesquite taught me many lessons, but two of them remain the most emotionally connected to me. First, I learned the difference between pruning something for its own good, following its natural form and inclinations, and pruning something with no purpose in mind other than retribution. This, I decided, was the difference between discipline and punishment: one looks forward and works toward restoration and health; the other looks backward and tears down, dehumanizes. Pruning should enhance life, not maim it. It was ironic how prison treated its plants and inmates in the same way.

Second, I learned that stubbornness doesn't mean selfishness. The mesquite was the only one of its kind at Santa Rita. It was alone but not withdrawn into itself. Cicadas trilled among it branches. Bees carted off its pollen and nectar. Ground squirrels feasted on its sweet fruit. The tree seemed obstinate only in the way that all life, by definition, is obstinate. It *wanted* to live. Despite the buildings and fences that confined it, the concrete slabs beneath it, despite even the brutal punishment of its careless pruning, the mesquite emerged each spring with offerings of leaves and shade, pollen and seeds. This was what struck me: The mesquite didn't shrink from this place; it made the best of it.

For Tony, nature, in the form of a pair of toads, meant remorse. For David, nature connected him to shame, perhaps, and a longing for forgiveness. For Steve, it was peace of mind.

For Brad and Byron, nature meant something else. Brad counted the barn swallows. Every season he recorded their numbers—eggs, days to hatching, days to fledging, successive broods, successful fledgings—since that first spring when the birds daubed a mud-pellet nest at the end of his run. He was dying, had been dying for twenty years but was getting closer when I knew him. His cancer and his colostomy bag, however, couldn't dispel the simple joy his birds brought to him on dark, narrow wings.

Then there was Byron who, for four years, kept a stunted ground squirrel in his cell after its mother had abandoned it. Byron adopted the runt, feeding it apple slices and carrying it around the prison yard in his pocket. He had "house trained" the squirrel to use a newspaper-lined litter box and to hide from the guards when they searched his cell for contraband like illegal drugs or pets. He could tell by the condition of its fur whether his companion was hot or cold. He could sympathize with it whenever it climbed to his window, stood on its hind legs, and called to others of its kind that never answered.

For Brad and for Byron, these animals allowed them to express feelings toward something other than themselves, and in an environment that promoted selfishness as a necessary survival skill, this kind of affection was a contradiction.

Another contradiction I saw in prison came in the form of a female mallard duck that nested every year inside the perimeter fences in a clump of umbrella plants. The first year, an officer discovered her and smashed her eggs with a rake. He called the nest of a dozen or so creamy white eggs a security risk. A security risk?! In subsequent years, however, the men kept secret her spring arrival, protecting her from guards with rakes, and bringing her slices of wheat bread and cups of water. At hatching, the female and her awkward, bicolored ducklings, escorted by her guardians, walked across the prison yard toward our waste-water treatment pond. At any time during her brooding, the mallard and her eggs might have become a casserole in someone's hotpot. Instead, the inmates made prison her sanctuary.

Prisoners and nature. It may be farfetched to suggest that nature can teach something valuable to prisoners. But prison, with its sadistic officers, its administrators bent on absolute control at all costs (financial and human), its medical personnel who have little regard for life, certainly can't instill anything meaningful. Unfortunately, these people are the prisoner's most visible role models. I would rather learn from nature than learn from prison. I would rather be a disciple of saguaros and centipedes. Nature cuts through more hard layers than brutality ever will, ever could. For me, nature touched the place where hope lay, and hope was a better security system than all the guards and fences and electric locks because hope moved me to believe in a better future. For twelve years hope was my jailer.

In my last years at Santa Rita, nature became a problem. The peppertrees in the inmate park had grown too tall. Someone could have climbed one, hid among the branches, and imagined he'd escaped. The rosebushes and mint had grown too green and lush. Someone could've lain down and disappeared into their leaves. Crews with a chain saw and backhoe worked feverishly to correct the error in security: All the roses, privets, the Texas ranger, all the Mexican bird of paradise, the desert willow had to go—had to be cut down, chopped into sections, wrenched from the ground. The few peppertrees that survived the clearing couldn't take much more pruning. They lacked all lower branches, their skinny trunks winding comically into high, tight crowns like trees in a Dr. Seuss story.

Then the policy of nature-as-risk spread its infection to the yards. Already, concrete had entombed flower bulbs and toads. Now, the landscape crews tore out the ocotillo, barrel cacti, and agave. Next came the yucca, the catclaw acacia, the New Mexican locust in front of my cell. Shrubs were the greatest danger to security; they all went. And more trees . . .

If I were confined now at the Santa Rita Unit, I couldn't connect to nature as easily as I could then. The prison is a different place—and not just because of the continuing landscape devastation as pea gravel replaces even the grass. Nature still invades there, by wind and storm and bird. What's different is that the men can't leave their cells; they're locked up all day and all night as the compulsion for security overrides the privilege of sanity. The fortunate ones find cockroaches in their cells.

Cut off nature from anyone and you cut off all possibility—something profoundly more inhuman than taking away his freedom. I can't imagine doing time there now—not experiencing a daily renewal in wildness, not learning about survival from a twisted and obstinate mesquite.

Note

This chapter is adapted from *Wilderness and Razor Wire: A Naturalist's Observations from Prison.* San Francisco: Mercury House, 2000.

THe nature OF MY LIFe

JAMES J. FARRELL

In a lot of writing about men and nature, the relationship seems remarkably close. Men engage nature in camping, backpacking, hiking, trekking, hunting, fishing, canoeing, kayaking, and so on. The encounter involves shooting nature or hooking it, mastering nature or just surviving it. It involves conquests of nature, and companionship in nature. At its best, it also describes immersion in nature and identification with it.

Our focus on these activities, however, depends on narrow definitions of both "man" and "nature"—or at least of masculinity and nature. When we tell these stories about men and nature, we often mean "real men" in "wild nature"—or *some* men in *some* nature. So these conventional definitions of "men" and "nature" have, I think, obscured more complex and potent relationships. For most American guys in the twenty-first century, these outdoor activities are pastimes, not our primary occupations. We may derive a substantial part of our identity from these pursuits of happiness, but our bumper stickers tell the real story. "I'd rather be fishing," we say, because most of the time we're not fishing. Writing about "real men" in "wild nature" often produces insightful and lyrical prose, but it doesn't generally help real men live responsible lives where we actually are.

Some of the confusion comes from what it means to be a man. From the time parents dress babies in pink and blue, praising one toddler's looks and another's strength, we're teaching children to grow up gendered. By the time we're adults, most guys know how to "act like a man." For most guys, masculinity is so ingrained that we don't need to think about it.

We're forced to think about our masculinity when conversations—including the conversations of popular culture—turn to the topic of "real men." "Real men," we were told years ago, "don't eat quiche." "Real men"—who are constitutionally tough—don't cry. "Real men" hunt and fish. They work hard and they play hard, maintaining a spirit of fierce competition in both arenas of their lives. "Real men" play contact sports, and now they play "extreme" sports too. They drink beer and eat man-sized portions of meat. They drive pick-up trucks, and they pick up chicks.

Well, as my kids say to me, "Get real!" The peculiar thing about "real men" is that there aren't really very many of them, because "real men" act like most men don't. By my count, "unreal men" outnumber the "real" ones by a large margin, so it could be that what *most* men do *is* what's real and masculine. If I'm real—and other men like me are real too—then it might be a good idea to see how most American men encounter nature most of the time, and to consider the implications of that encounter for masculinity and for the natural world.

After all, we are in nature even when we're not in the boat or the blind, in the tent or on the trail. Guys like me spend more time inside than outside, and our relationships with nature are more mediated—in multiple senses of the word—than our ancestors' experience of nature. But we're never really out of nature. We're inside nature even when we're inside. Our language makes it difficult to think about this nature of inside, and especially about a man's relationship to nature at home. In this culture, we celebrate the "outdoorsman," but we don't even have a word for the hearty "indoorsman," the man who communes with nature behind closed doors. There's a publication (mostly for men) called *Outdoor Life* but no comparable men's magazine called *Indoor Life*.

We live in spaces filled with nature, but we don't always see the forest for the trees. In this essay, I propose to explore the nature we occupy day-to-day instead of the one we inhabit on weekends, on vacation, and in our fantasies. This essay looks at the nature of my daily life, which is not the nature of Thoreau or John Muir or Edward Abbey. It's the nature of suburbia, the nature of home, the nature of television. It's the nature of houseplants and pets, Kleenex boxes and air fresheners. It's watching nature on the weather report, in specials on PBS, and in the ads for SUVs. It's complex, and it's contradictory. It's still a gendered experience of nature, but in real life, gender differences aren't as important as gender similarities. Still, it might be useful to think about how to act like a man in the nature of my everyday life.

In doing this, I'd like to speak not as an expert in masculinity or the natural world but as a guy willing to think twice about the patterns of his life. I'd like to look at a few common things and see what they tell us about the common sense of our culture. Sometimes this will result in what Elizabeth Minnich calls "blinding flashes of the obvious." In other cases, it may result in blinding flashes of the oblivious—when I don't see things that other people do. In the process, I'd like to be suggestive but not conclusive. I think of an essay not as a monologue on paper but as part of a conversation. So I hope you'll join in.

JAMES J. FARRELL

Day-to-Day Nature

I'd like to suggest that we experience nature in three ways in our daily lives—as representative nature, as domesticated nature, and as instrumental nature. Throughout our homes, we encounter representations that place us in a peculiar relationship with nature. Landscapes, flowers, and air fresheners, for example, offer idealized images of nature that reflect and affect the real lives we lead. In our homes, too, we often experience domesticated nature. Both houseplants and pets offer interpretations of nature and culture and teach us about our relationship with nature. And finally, instrumental nature—the natural stuff that makes cultural stuff work—powers the lives we lead, even indoors. These three natures—obviously overlapping—offer some insights into the nature of our lives.

In many American homes, nature functions as representation, and these ornaments speak to issues of nature and culture, and to masculine and feminine cultures. American homes are full of representations of nature. This mediated nature literally offers a sense of nature—usually a visual sense, but sometimes a sense of smell or sound. The picture over the sofa is a pastoral scene with a covered bridge. The Kleenex box in the bathroom is festooned with flowers. A spray of colorful blooms highlights the dinner table, while dried flowers and silk flowers—and even plastic flowers—grow in other pots and baskets.

The most popular paintings in American homes are landscapes. For some reason, we like the look of nature on our walls. And the landscapes we like are a particular kind. We like scenes of tranquil nature, of the pastoral and picturesque. At home, in our art, we don't appreciate stormy weather or the impenetrable tangles of the jungle. And we don't seem to like people in our pictures. The landscapes we like are landscapes we don't inhabit, so we populate our homes with depopulated landscapes. The landscapes we hang over the sofa seem to serve as a counterpoint to the lives we live in our living rooms (Halle 59–86).

A second popular genre of representative nature is floral. Many of our interior decorations are attempts to "say it with flowers." In intimate relationships, we learn to express our romantic love by speaking the language of flowers. Of course, the flowers on the Kleenex box or on the living room upholstery aren't a form of romantic love. But they are a sign of a nature lover, perhaps even a form of biophilia. We plant these flowers on the surfaces of our lives to express our affection and affiliation with nature. We use them to show our appreciation for the profligate beauty of the natural world, and for the artistic ability to reproduce—and sometimes enhance—it. We also use them to smooth out the seasons, so that we can have nature's blooms in our homes even when there are no blooms outside our windows. Even though flowers seem to express a general affection for nature, they express a particular affection for nature's ornamental effects, for the flowers rather than the fruits of nature.

In a similar fashion, we use air fresheners to express our appreciation of nature's scents. The smell market used to be defined by these functional products that masked smells. But these days, manufacturers make most of their dollars and cents by making scents that function differently—as nasal narratives. Glade, for example, comes in thirty-eight different aromas, ranging from fruity to floral, from a Mystical Garden to a Hawaiian Breeze. Air fresheners like Glade are now a lifestyle product used to tell stories by making smells. Its stories usually evoke natural places: Country Garden™, for example, "reminds you of sitting on a swing with your head thrown back and the warm sunlight striking your face. Lift your feet from the ground and let yourself soar away." Botanical Garden® is "a perfect combination for rejuvenating your body, mind, and soul." These scents aren't just for the nose, they're for the imagination. They're aromatherapy without the therapist, based on the assumption that smelling good is a path to feeling good.[1]

The efflorescence of the home—the flowery patterns and the floral smell—is culturally feminine. These air fresheners are marketed mainly to women, because women shop for household products, and because women pay more attention to the sense of smell. In the gender divisions of American culture, men are supposed to smell, and women are supposed to smell good. Perfumes are primarily for women, although men can be enticed to apply scents that are called something else—cologne, or conditioner, or skin bracer.

Many men appreciate the natural touches of these landscapes and flowers and scents, but guys have distinctive ornamental traditions too. Often a den or family room will be configured as a man's space, with wet bar, pool table, and—if the man is a sportsman—trophies of the hunt. The deer's head or the walleye is mounted on the wall to show the beauties of nature and the prowess of the hunter or fisherman. Men also use colors to mark their domestic territory—with carpets and accessories in darker colors like "hunter green" or "navy blue" instead of the pastels that usually predominate in the rest of the house.

In their dens, many men also use wildlife art to express a relationship with nature. Scenes of animals in the wild are prized both for their realism and for their symbolic associations—including associations with masculinity. Wildlife art often offers a view of nature from the hunter's eye—not just the eye in the gun sight but the aesthetic eye, the eye that appreciates the beauties of nature that are an underlying element of the hunting experience. Wildlife art uses the realism of details to present fantasies of nature. The painting of ducks rising from a sun-soaked pond offers a portrayal of plenitude that's deeply satisfying. Often, wildlife artists—like Minnesota's Terry Redlin—are associated with conservation organizations, hoping that the realism of their art can counteract some of the realities of the modern world. Like other instances of ornamental art, wildlife art offers us the satisfactions of a natural world that's undermined by the everyday world we live in (Russell 45–49).

A final form of representative nature—at least in this essay—is the nature that's boxed and presented on TV. As couch potatoes, we naturally root ourselves in our furniture to watch mass-mediated images of nature. Although we don't generally identify with televised nature the way we do with hunting or fishing or backpacking, some of our most significant encounters with nature occur in prime time. On a daily basis, for example, most of us see more wild nature on TV than we see outside. Traveling by remote control, we venture into the wilderness of American assumptions about nature.

Indeed, several cable channels specialize in nature, offering instructive entertainment. National Geographic, Animal Planet, and the Discovery Channel offer programming that informs us about the nature of our world. But as Bill McKibben suggests in *The Age of Missing Information*, this programming also in-forms us—or forms us within—by its concentration on certain themes and its omission of others. Nature programming depends, for example, on good visual images and good stories, so it tends to concentrate on colorful creatures that can be anthropomorphized—the same species, as it happens, that can be plushed into Beanie Babies. TV likes action, and so we see the "highlight films" of natural life: we don't see lions sleeping, which is what they mostly do; we may see them sitting majestically, but more often we see them hunting, or prowling, or tending the young. Like wildlife art, nature programming often tries to raise our awareness of endangered species and extinctions, but these rare plants and animals appear on our screens so regularly that they seem plentiful. "The upshot of nature education by television," says McKibben, "is a deep fondness for certain species and a deep lack of understanding of systems, or the policies that destroy those systems" (78–79).

On the broadcast channels, nature is more evident in ads than in programming, which usually focuses on the indoor adventures of *Homo sapiens*.

The nature of TV advertising is often gendered, but in advertising for both men and women, nature is most often presented as a curative or a counterpoint to contemporary living. In ads for women, nature tends to be pastoral or picturesque, beautiful and comforting. Instead of assuming that a woman surveys the landscape, there's often an assumption that a woman is surveyed *in* the landscape or *as* a landscape. Women are more often naked in nature. While men are often dressed and doing something *in* nature or *to* nature, women are often undressed and doing nothing. Although they may be contemplative, they're more passive, and they seem at times to be a part *of* nature. In perfume ads, a dreamy woman often appears in a dreamy romantic landscape. In some way, because she has applied this scent to her body, she becomes one with nature—and often with the man of her dreams. The dreaminess of such ads is intentional, in part because you'd have to be half-asleep to fall for the loose connections. But it's clear that advertisers want us to equate the liquid in the bottle with the purity and beauty of the natural world. So women who apply these "natural" scents are also applying a long tradition of landscape aesthetics.

In ads for men, nature tends to be distant and sublime, remote and challenging. In this advertising, nature is often where "real men" go to get away from civilization and from the civilizing women who stultify our manliness. Ads for SUVs especially seem fueled by testosterone dreams, evoking themes of escape and adventure, the sublime and the subliminal. In SUV ads—if not in SUVs themselves—we can escape the clutter and congestion, the rules and responsibilities, of our everyday lives. "You would not be very human," says one ad, "if you did not mostly prefer to be not-here, not-now." So we light out for the territory. Driving an SUV—a Pathfinder or Blazer or Explorer or Expedition—we feel magically connected to the pioneer spirit and conquest mentality that made this country great by making it a man's world. We engage in the new "technological primitivism," whereby Americans invent increasingly intricate technologies to take them back to the places technologies took them away from.[2]

Those places are both geographic and chronological. SUV ads are often nostalgically set in the landscapes of nineteenth-century painting and photography—in deserts, on mountains, in the uninhabited spaces of the American sublime. These scenes of "Marlboro country" appeal to our aesthetic sensibilities, our genuine amazement at the beauties of nature. Market research shows that they also appeal to the values and lifestyles of truck drivers, who think of themselves as both nonconformist and environmentally conscious. And, as you might expect on tele*vision*, they appeal mainly to the eye.

But there's also something subliminal about this "SUV sublime." For all of the purple mountains' majesty, there's a sense that the driver (usually male) is really king of the road (and the off-road too)—and king of the planet. The nineteenth-century sublime was intended to evoke humility, but the "SUV sublime" evokes hubris. Along with the ads promising that an SUV will help you conquer any kind of weather, these ads show how to "Control your Mother [Nature]." Ford promises us a world with "no boundaries," and they don't just mean geographic boundaries. They mean no bounds to human power, and no limits to traditional "kick-ass" masculinity.

Of course, these TV ads are pumping irony by using nature to sell machines that degrade nature, whether or not they ever go off-road. The ads show SUVs splattering through muddy creeks, screaming along dirt roads, and sliding in the desert sands. Dirt seems to be an essential attraction of these vehicles, reassuring men that we can immerse ourselves in nature by driving through it. But even on the suburban street or in the Wal-Mart parking lot, the emissions from these behemoths affect the atmosphere, and—eventually—the climate of American life.

Watching TV, we imbibe the images and assumptions of our culture, including our dominant assumptions about nature. As with other instances of representative nature, TV reproduces the conventional sights and sounds and smells of nature in our homes. Still, because somebody else usually designs this simulated nature, our relationship with representative nature is remote and controlled. With domesticated nature, however, we interact with the natural organisms themselves.

At home, the nature we usually encounter is thoroughly domesticated, consisting primarily of pets and houseplants. Houseplants sprout in pots on shelves and stands, in corners and hallways. Spider plants cascade in the windows. And pets poke their noses into our lives at every turn. Pets and houseplants are primary teachers about the natural world, and most American children get their first lessons about nature from the domesticated strains that populate the ecological niche we call home. So these days, when boys are learning to be men, this is the nature they encounter.

Like most American families, Barb and I have a few houseplants in our house. We grow our houseplants not for food or for feed—even though pets and errant children do sometimes nibble them—but for a high yield of meaning. Even though we don't eat them, our plants say that we have taste. They also tell us that we have a nurturing relationship with the natural world. Houseplants are, when you think about it, a relatively recent phenomenon. Human beings haven't always kept houseplants, since people historically spent most of their time among plants outside. But as we've increasingly planted ourselves indoors, we Americans have apparently felt a powerful need for the nature we've walled out. We need nature, it seems, and not just abstractly. So now, instead of going out where the plants are, we bring them in where we are.

Houseplants also tell us that we value nature, even though we're members of a society involved in the most sweeping natural destruction in world history. With houseplants, we can value nature, not as a system but as a sign, not as land but as landscape. We pose our plants in relation to the rest of our lives, and by doing so, we show that we like the look and feel of nature; and we especially like the *look* of nature under human control. There aren't many people, for example, who choose ugly or aggressive breeds like crabgrass or creeping Charlie for their houseplants.

Houseplants tend to be women's work. This is because the care of houseplants is a form of housework, and in America, housework is still disproportionately women's work. At the same time, houseplants are decorative and ornamental, and we still tend to think of women as more inclined to decoration and aesthetics than men. So women do most of the work with houseplants, but men share in the harvest of meanings.

For millions of Americans, pets are another common way to come in contact with the natural world. Domestication is a process whereby human beings make nature cultural; etymologically, it's how we convert nature to household uses. In some cases, it's how we make nature feel at home, since some animals are more domesticated than others and actually get invited into the house. Sixty-three million American households own pets. Dogs live in 40 million households, with cats in 34 million. After that, populations are smaller (both numerically and physically). Freshwater fish, birds, and small animals come next, followed by reptiles and saltwater fish.

These animals are male and female, but the American menagerie is also masculine and feminine—not according to the pets' proclivities but according to humans' cultural coding. Dogs are more masculine than cats, and big dogs are more macho than small dogs. Hunting dogs are more masculine than show dogs, and predators are more masculine than herbivores. Even though Barb and I describe our bunny as an "attack rabbit," it's not an animal that makes a man out of me.

When American pet owners take the dog for a walk or sit down to stroke the purring cat, we don't usually imagine ourselves getting back to nature. Ironically, when we most feel at home with nature, we don't imagine it as nature at all. This is because we still imagine nature as wild, untouched by human hands (and therefore unavailable to humans, most of whom have hands). When nature is modified by a human touch, somehow it becomes less natural (Cronon).

People aren't pets, but we too *are* domesticated animals. In our culture, however, when we think of "human nature," we seem to pay more attention to "human" than to "nature." We talk about people and animals as if these were mutually exclusive categories. Still, unless you categorically define people as not-nature—and many of us implicitly do—there's not a lot of evidence that people are supernatural. Large mammals with big brains, we have evolved into a form of natural life conscious enough to reject a definition of ourselves as nature. We may transcend nature, but we do it with bodies and brains that have evolved in nature. So when we encounter our partners and children and friends at home, and even when we communicate electronically with each other in cyberspace, we're encountering nature too.

In America, our homes are the places where we care for this human nature—physical, psychological, sometimes even spiritual. And as with houseplants and pets, women seem to do most of this natural work. They deal with most of the nature of human nature—the inputs and outputs of physical beings. They bear children and perform the lioness's share of the childcare. They change the diapers, wipe the snot, mop the vomit, and deal with other excretions and expectorations of the human young. Women still prepare most of our food and serve most of our drinks. If we acknowledge our embodied existence, our identity as animals, then women seem to be dealing most directly with it. In the nature of American daily life, women seem to know more than men about relating to nature where they are, instead of someplace else.

With pets and houseplants (and people), we see nature in the house. But a lot of housekeeping is devoted to keeping other intrusions of nature out of the house. Our houses have doors to admit people, and windows to admit light. But we also have doors and windows—and screens and filters—to keep out snakes and rodents, bugs and bees. And we have all sorts of cleaning tools to suck up the nature—in the form of dirt and dust—that infiltrates the home. In many ways, housekeeping is the art of keeping this invasive nature out of sight and out of mind. As Suellen Hoy suggests in *Chasing Dirt*, a lot of household chores involve keeping nature in its place.

This also involves keeping men and women in their places. Men do more housework than ever, but women still perform most of these essential cultural rituals. A recent survey shows that men and women share responsibility for most household tasks, but women do most of the cleaning and laundry, while men dominate in taking out the trash, cleaning the porch, maintaining cars and household appliances, mowing the lawn, painting the house, and cleaning the gutters. Men fight the entropy of the house and household appliances, while women fight dirt and germs. Women care for the social connections of the household, while men often take care of its business connections.

Men usually mind nature outside, and in transitional tasks like taking the trash out. Garbage is what we might call residual nature—what's left after we've used a natural resource. So it's one way of understanding our relation to the natural world. Americans make garbage at home; it is one of the few remaining forms of home production. Each week, we fill our houses with materials from the supermarket and the mail and the mall, and each week, with ritual regularity, we discharge it into the waste stream. The average American manufactures between two and four pounds of garbage a day; in a year's time, for a family of four, that's three or four tons.

Another household task that falls mainly to men is yard work. During the summer, for example, I commune with nature once a week behind my lawnmower. I take a hike back and forth across the lot, grooming the trail as I go. A lawn is a farm in which the main crop is three or four grasses harvested weekly and then thrown away. So keeping a lawn is a species of agriculture in which the main crop always costs more to produce than it's worth. It is a sign of conspicuous consumption, an ornament to affluence, a natural wonder that says that the proprietor can afford its uselessness.

Like many forms of domesticated nature, lawns are an important natural element in an American culture of distraction. Every minute that I spend thinking about the nature of my lawn is a minute not thinking about something important. Every hour spent mowing and clipping is an hour not devoted to environmental restoration or the common good. Indeed, instead of grass-roots politics, Americans increasingly have grass-roots inaction, as cynicism and skepticism are fertilized and watered by politicians, the press, and talk-show hosts. In a world that seems out of control, we turn to the manageable turf of our own backyard.

From houseplants to pets, from garbage to garage to lawn, American men and women encounter domesticated nature in our daily lives. Representative nature displays many of our presuppositions about nature—or at least about the nature that we have in mind. Domesticated nature embodies many of those presuppositions. And instrumental nature calls them into question. When we *use* nature at home, we're often engaged in activities that contradict the feelings of nature we express in our decorations and domesticated relationships.

Representative nature offers images for me to contemplate. Domesticated nature gives me a way to care for some nature. But instrumental nature helps me get things done. Gravity, for example, is invaluable. It helps me keep my feet on the ground, and it keeps me in my seat when I'm writing. But it's also responsible for the water pressure that powers my shower and my toilet and that provides liquid for my coffee every morning. The water itself flows from the faucet when I turn the tap, and it stops when the valve closes again. But the pressure of a million gallons of water in the tower near my home keeps water at my command.

At home, which is where most of our interactions with nature occur, femininity and masculinity have significant meanings, but when it comes to instrumental nature, they don't mean much. When I take a shower, the interaction with nature is pretty much the same as when my wife takes a shower. We're both relying on city water, pumped from the Jordan aquifer, purified with chemicals, pumped again to water towers throughout the city, and flowing thereafter to the sinks, bathtubs, washing machines, dishwashers, and sprinklers of suburban America. Both Barb and I count on the gas company—and our water heater—to heat the shower water for us. And we each use a bar of soap and a bottle of chemicals to clean our body and its hair. Naturally (culturally), her shampoo smells different from mine.

Electricity also works for me twenty-four hours a day. Right now, for example, I'm sitting in front of my computer. It powers up with a whirr and a collection of clicks when I turn it on in the morning, and it stands by in anticipation when I'm reading elsewhere in the house. While I write, I listen to the radio, using electricity to generate a musical stimulus. Electricity is the genie that heats my food on the stove and in the microwave, and it's the magic that powers the icebox that hums with my frozen food and refrigerated leftovers. At night, the lights in the house are second nature to me, as are the images that dance on my television. Like most Americans, I'm wired—and basically, I'm wired to nature.

In Minnesota, the energy that powers my computer comes from coal and nuclear plants and ultimately from nature outside the state. I don't burn coal in my fireplace grate; it's burned for me at the Sherco plant in Becker, Minnesota, where Xcel Energy burns thirty thousand tons of coal a day. Sherco uses low-sulfur coal from Montana and Wyoming, but its emissions of carbon dioxide still make "CO_2AL," what the *Economist* calls "environmental enemy No. 1." By the time it reaches me, however, the coal has been transformed into invisible electricity, so I tend to think of it as clean energy (Xcel Energy 11).

Some of our most intimate encounters with the natural world come at the dinner table, when we commune with nature by eating it. In fact, if we didn't eat nature, we would die. But we don't generally think of eating a meal as getting back to nature. We consume the body and blood of animals, and the leaves and flowers and kernels and seeds of plants. Our bodies, which are nature, are nurtured by the nature we culturally define as food. As Wendell Berry suggests, eating is a natural act. Whether we eat tofu or Tater Tots, the food that enters our lips connects us to nature and makes us responsible for it. As Berry says, people who eat food "must understand that eating takes place inescapably in the world, that it is inescapably an agricultural act, and that how we eat determines, to a considerable extent, how the world is used." As the food goes into our mouth, the institutional food chain is activated to restock the supermarket shelves. In consuming food, we nourish both ourselves and the institutions that provide food to us ("The Pleasures of Eating").

At the other end of things, we dispose of our bodily wastes in the bathroom. Our toilets are machines that use water to wash wastes away. They are the primary example of what sociologist Philip Slater calls "the Toilet Assumption" of American life. He describes this principle as "the notion that unwanted matter, unwanted difficulties, unwanted complexities and obstacles will disappear if they are removed from our immediate field of vision." Incrementally and excrementally, he suggests, Americans have segregated the ugly and the beautiful, pain and pleasure, them and us. The result of such efforts, Slater says, is "to remove the underlying problems of our society farther and farther from daily experience and daily consciousness, and hence to decrease, in the mass of the population, the knowledge, skill, resources and motivation necessary to deal with them." In this way, he suggests, our restrooms are connected to a society of structural *un*-rest (15–19).

Instrumental nature confounds many of our accounts of men in nature. When we commonly tell stories about men and nature, we're usually talking about direct experiences of men and the culturally "natural" world. But most of our experience these days is indirect. We affect nature most when we're thinking about it least, because most of our interactions occur by remote control. We understand the remote control for the TV, but we forget that the faucet is a remote control too. The electric switch is a remote control, as is the keypad on the microwave. The thermostat operates the furnace and the air conditioner, but it also activates the remote flow of gas wells in other parts of the United States. Both the lawn mower's starter and my car keys are remote controls, affecting the pumping of oil and the international politics of the Middle East. A masculinity that doesn't account for these incursions in nature isn't accounting for much.

Men usually construct America's architecture. Although women are entering the construction trades, it's still largely a "man's work," so it's usually a crew of guys who hammer up a house. It's guys who install the pipes and ducts and cables that conduct water and gas and electricity into the house and let sewage and waste water and stale air out. It's still usually guys who come to service the furnace or the dishwasher or the dryer. So, in our capacity as workers, we men construct the Great Indoors. Once we physically construct it, however, it seems to be socially constructed as women's sphere. Both interior decoration and housework are—relatively speaking—given over to women, so ornamental nature and domesticated nature are often seen as part of a woman's world. And there aren't a lot of "manly" things we do in this Great Indoors.

I'm a man, and my life depends on nature. But in most of my life, neither my masculinity nor my natural qualities seem very prominent. When I come inside, it seems, I'm supposed to check my masculinity at the door. Indoors, many of my interactions with nature are symbolic or domesticated. But most of them are instrumental, mediated by the institutions and infrastructure of American life. Flipping switches, turning faucets, pressing the "power" buttons on my tools and appliances, I'm blundering around in nature without really noticing it.

American men and women seem to have adapted to the nature of home by carving out roles and rules for behavior. But neither men nor women have adapted very well, because the nature of the American home is an environmental catastrophe. It's important to see the gendered ways of relating to nature. But it's even more important to see our equal-opportunity oblivion, caused by the consumer forgetfulness of instrumental nature. Farmer and poet Wendell Berry, for example, claims that, "with its array of gadgets and machines, all powered by energies that are destructive of land or air or water, and connected to work, market, school, recreation, etc., by gasoline engines, the modern home is a veritable factory of waste and destruction." Even worse, he claims, the modern home distances us from our destruction (*Unsettling of America* 51–52).

As Berry suggests, the nature of my life is that I don't usually see the nature in it. The nature of my life is that my conventional masculinity isn't an appropriate response to the needs of nature (and of human nature) in our time. Modern masculinity isn't adequate to men because it's not useful in enough circumstances.

This gives us some choices. We can continue to leave the nature of our lives out of sight and out of mind. Or we can begin to practice a discipline of seeing through the literal superficiality of our lives. Like Superman, we men need a kind of X-ray vision to see beneath the surfaces of American culture, to give us a sight of the unsightly behavior that makes our lives work, and of the *natural* foundations of American culture. We can begin to *see* and *feel* nature at the end of the food chain, the product chain, the gasoline hose. Without substantive changes in the ways we see and treat nature at home, we're likely to become a nation of nature-loving nature-killers.

These new ways of seeing will involve new methods of socialization. Americans desperately need a new kind of "naturalization" process. Right now, the term "naturalization" refers to the process of socializing people for citizenship. After learning about the United States and its democratic institutions, foreign-born people make a commitment to their new country. We need a deeper sort of naturalization process. We need to teach ourselves—and our children—to love nature, and not just the nature we "go back" to. It's great to love the nature of the national parks and the wilderness, but we also need to love the nature flowing through the faucet and the toilet, the wires and pipes of the domestic infrastructure. After learning about our complicity *as* nature *in* nature, we need to make a commitment to this complex New World with a declaration of interdependence (Cronon 88–90).

We also have choices about our masculinity. We can cling to a largely irrelevant masculinity, acting manly mainly in the extraordinary moments of our lives. We can continue to define masculinity as we did in middle school—as "not feminine." Or we can decide to shape our masculinity to our ordinary circumstances, in lives we share with women and children. We can complain that we've been "emasculated" in the modern world, or we can begin to "masculate" the world we really live in by applying some of our traditional masculine values to our present circumstances.

To begin with, we need a modern masculinity that's more extensive and inclusive. Our traditional masculinity tends to encourage exclusivity. It's been a way of dividing the ("real") men from the boys, and—even more important—the men from the women. We're put down for running or throwing "like a girl," and we're labeled "sissies" or "pansies." Many of our most conventional stories of men in nature involve substituting a relationship with Mother Nature for relationships with our mothers and sisters. Perhaps it's no coincidence that Minnesota's fishing opener falls so often on Mother's Day weekend. In many ways, traditional masculinity keeps us away from women, and away from home.

Obviously, many men who have found this masculinity restrictive have carved out good lives by bending gender roles. Guys cook and clean and care for their kids. Guys can be sentimental and show their feminine side. As women move into "a man's world," men are increasingly able to move into a woman's world. But they don't get the approval of the masculinity cops. Men can be different, but the difference is always marked as a deviation from a fairly narrow norm.

I think it's time for men to share in the housework—and not just at home. Both men and women will need to pay attention to the housework that bridges home and the world, to the environmental impacts of household decisions. Both "ecology" and "economy" derive from the Greek word for "household," and we'll need to honor that etymology by undertaking the housework of maintaining the planet.

Thoreau said that "in wildness is the preservation of the world." As usual, he was about half right. These days, it also seems true that "in domestication is the preservation of the world." As we've seen, even indoors, Americans regularly and ritually express their affinity for nature. But as with so many family values, this affection seems to be confined to the private sphere. We seem to love our houseplants and housepets more than we love the ecosystems that provide the natural resources for our profligate lives. In the twenty-first century, we need to extend our domestic values of care and cooperation to what Gary Snyder calls "Earth/House/Hold."

There's a lot of writing about ecofeminism, and most of it is right on target. Western cultures—patriarchal cultures—have socially constructed relationships between men and women and between men and nature that are often pathological. But while patriarchy is problematic, not all men—or even all masculinity—is equally problematic. It might be useful, therefore, to think about what ecomasculinism might look like—to think about the natural and cultural resources we might have for this comprehensive housework. We have, I think, traditional masculine values that do, in fact, have applications (for both men and women) at home.

The masculine role of provider, for example, provides a possible framework for a more responsive and responsible relationship with nature. Instead of just providing for our families, men will need to provide for the nature in our families' lives. We can't just take from the earth to provide for the family, because that impairs our ability to provide for the family's future. But our families need to become more inclusive too. The earth needs to *become* family, and to be provided for too. So paradoxically, we might find that, as men, we provide more for our family by providing less.

In the past, men have prided themselves on their abilities as protectors, especially as protectors of women and children. But we can't protect our social systems like family, community, and nation without protecting the ecosystems that support us. For ecomasculine men, the department of defense will include defense of the biosphere. John Wayne is dead, bless his swaggering soul. But like John Wayne, we can protect the earth from hostile forces—and especially from the hostile forces in ourselves.

Men have also traditionally asserted their masculinity in their mastery. In the past, mastery meant imposing one's will on other people and creation, either by force or by force of will. In the future, mastery will likely be achieved by aligning one's will with other people and the creation. And that's what men need to do now—take control of the extinctions and extractions that aren't visible on our television screens. Men who are in charge of the remote control will need to take control of their remote control of the earth and its ecosystems.

Men have traditionally prided themselves on their competitiveness and their sportsmanship. But win/lose competitiveness will need to be supplemented by a competition in which everybody benefits. Following the example of sportsmen's organizations that have practiced conservation to preserve sportsmanship, men will need to help their families and friends to practice conservation to preserve the earth.

Toughness has been the heart of a man's masculinity. In the new era, men will need to be tough enough to be soft. We'll need to admit—rationally, psychologically, emotionally—the hard truth that, in an ecological age, we're vulnerable. But like tough athletes we admire—like Lou Gehrig or Cal Ripken—we'll need to play through the pain, or at least the discomforts, that will accompany a shift to a sustainable society.

Historically, a real man has been willing to take responsibility for his actions. In our present situation, a man's man will take responsibility for his actions, both intentional and incidental, both visible and invisible. This heightened responsibility will entail a more complex understanding of environmental impacts, both positive and negative, and a willingness to understand ourselves as actors in systems that are as large as a planet.

Even though men have been taught to be responsible, they've sometimes exercised that responsibility by lighting out for the territory. The desire to escape can be a reasonable motivation, especially when civilization stunts the growth of human beings. It's a good sign that we recognize the dehumanizing aspects of our society. But a real man might consider fight instead of flight, changing the civilization instead of changing his location. He might consider constructive politics instead of apolitical apathy.

In the past, men have taken pride in their inventiveness. We've lionized inventors like Thomas Edison and Henry Ford and Steve Jobs for their ingenuity. We'll need such inventiveness in the future. We'll need people willing to tinker with nature without trifling with it. These ecological designers will follow nature's example to create regenerative designs that power both people and the planet. For centuries, Americans have believed that progress is our most important product, and we've often measured our progress by our inventions. But now, a new definition of progress will need to be the first product off the intellectual assembly line.

In the past, men have prided themselves on being rational and efficient. We see the big picture and leave the "detail work" to others. Such a division of labor won't work in the future, but such vision and thinking will still be essential (for men and women). The big picture will be bigger than a spreadsheet. Any government or business or family that thinks on a scale smaller than a biosphere will be thinking too small. And rationality and efficiency will be applied to cultural processes, as we try (finally) to adapt our economy to its environment, which is the whole earth. Rationality and efficiency stand opposed to the wastefulness of consumer culture.

As we face nature in the future, we'll be looking in the mirror. We'll need to be men, and we'll need our masculinity. But a new masculinity can't be one that separates us from ourselves, from females, or from the earth. Instead of proving our masculinity by taking on nature in the wild, we'll need to act like men right where we are.

Notes

1. See the Glade section of the S. C. Johnson Web site at www.glade.com.
2. For a new look at the SUV boom, see Keith Bradsher, *High and Mighty: SUVs—The World's Most Dangerous Vehicles and How They Got That Way* (New York: PublicAffairs, 2002).

Works Cited

Berry, Wendell. "The Pleasures of Eating" In *What Are People For?*, 145–52. Berkeley: North Point Press, 1990.
———. *The Unsettling of America*. San Francisco: Sierra Club Books, 1976.
Cronon, William. "The Trouble with Wilderness; or, Getting Back to the Wrong Nature." In *Uncommon Ground: Toward Reinventing Nature*, edited by William Cronon. New York: Norton, 1995.
Halle, David. *Inside Culture: Art and Class in the American Home.* Chicago: University of Chicago Press, 1993.
Hoy, Suellen. *Chasing Dirt: The American Pursuit of Cleanliness.* New York: Oxford University Press, 1995.
McKibben, Bill. *The Age of Missing Information.* New York: Random House, 1992.
Russell, David. "Wildlife Art on the Wing." *American Demographics* 10 (December 1988): 45–49.
Slater, Philip. *Pursuit of Loneliness: American Culture at the Breaking Point.* Boston: Beacon, 1970.
Xcel Energy. "Power-Generating Facilities—Minnesota," at www.xcelenergy.com/aboutUs/GenFacilitiesMN-Sherco.asp; "Environmental Enemy No. 1," *Economist* (6 July 2002): 11.

Notes on Contributors

MARK ALLISTER teaches English, American studies, and environmental studies courses at St. Olaf College in Northfield, Minnesota. He recently published *Refiguring the Map of Sorrow: Nature Writing and Autobiography*.

JAMES BARILLA is completing his doctoral degree at the University of California, Davis, where his research focuses on the intersections between ecological restoration and literature. A recipient of several research grants and awards, he was most recently a writer-in-residence at the Mesa Refuge, where he completed a nonfiction manuscript about fly-fishing across America.

RICK FAIRBANKS, dean and vice president for academic affairs at Northland College, has published articles on Nietzsche, the philosophy of science, and citizenship. He is an avid kayaker.

JAMES J. FARRELL, professor of history, American studies, and American conversations at St. Olaf College, is also known as "Dr. America," curator of the magnificent (but wholly imaginary) American Studies Museum, where he gives weekly radio tours on National Public Radio station WCAL. His books include *Inventing the American Way of Death, 1830–1920; The Nuclear Devil's Dictionary; The Spirit of the Sixties: Making Postwar Radicalism*; and, most recently, *Malls of America: Shopping for American Culture*.

CHERYLL GLOTFELTY is associate professor of literature and environment at the University of Nevada, Reno, where she cofounded the Literature and Environment graduate program in English. She coedited *The Ecocriticism Reader: Landmarks in Literary Ecology* and is currently editing an anthology of Nevada literature and writing scholarly essays on Nevada literature and culture.

LILACE MELLIN GUIGNARD has a master of fine arts degree from the University of California, Irvine, and is currently in the Literature and Environment program at the University of Nevada, Reno, where her research focuses on gender and nature. She has been a recipient of a Nevada Artist Fellowship, and her prose has appeared in *Whole Terrain*, the *Santa Monica Review*, *Orion Afield*, and the anthology *Young Wives' Tales*.

ALVIN HANDELMAN lives in Northfield, Minnesota, and is working on a novel about York, the slave who accompanied Lewis and Clark to the Pacific.

JIM HEYNEN has published widely as an essayist, poet, short story writer, and novelist. His most recent publications include *The Boys' House: New and Selected Stories; Standing Naked: New and Selected Poems;* and the novel *Cosmos Coyote and William the Nice.* He is writer-in-residence at St. Olaf College in Northfield, Minnesota.

KEN LAMBERTON's first book, *Wilderness and Razor Wire: A Naturalist's Observations from Prison*, won the John Burroughs Medal in 2000 for outstanding nature writing. His book about the natural history of the Chiricahua Mountains of Southeast Arizona was published by the University of Arizona Press in October 2003, and a second collection of prison essays called *A Sentence of Place* is forthcoming from University of Arizona Press in the fall of 2004. He lives with his wife and daughters in Tucson.

GRETCHEN LEGLER teaches creative writing, literature, and women's studies at the University of Maine at Farmington. Her essays about the natural world have appeared in venues including the *Women's Review of Books, Orion,* and the *Georgia Review.* Her collection of essays about Antarctica is forthcoming.

JULIA MARTIN lives in Muizenberg, South Africa, where she teaches English and creative writing at the University of the Western Cape. She has published stories and poetry as well as essays in the field of ecology, Buddhism, and literary studies.

STEPHEN J. MEXAL is a doctoral candidate in English at the University of Colorado. He has published on the spectacle of terrorism and is currently researching African American musical forms.

DAVID COPLAND MORRIS has published essays on Edward Abbey, Robinson Jeffers, John Muir, and the canon of American literature. He is an associate professor in the Interdisciplinary Arts and Sciences program at the University of Washington, Tacoma.

PATRICK D. MURPHY, professor and chair of the Department of English at the University of Central Florida, is the author of *Literature, Nature and Other: Ecofeminist Critiques; Farther Afield in the Study of Nature-Oriented Literature;* and *A Place for Wayfaring: The Poetry and Prose of Gary Snyder,* as well as editor of *The Literature of Nature: An International Sourcebook.*

SCOTT RUSSELL SANDERS is the author of *Hunting for Hope; Staying Put;* and *The Force of Spirit,* among other books. He teaches at Indiana University, in the beech-oak-hickory hill country of the White River Valley.

SCOTT SLOVIC is professor of literature and environment at the University of Nevada, Reno. He is the author, editor, and coeditor of numerous publications, including most recently *The ISLE Reader: Ecocriticism, 1993–2003* and *What's Nature Worth? Exploring Narrative Expressions of Environmental Values.*

THOMAS R. SMITH is a poet and essayist living in River Falls, Wisconsin. His books of poetry include *Keeping the Star* (1988); *Horse of Earth* (1994); and *The Dark Indigo Current* (2000). He was a founding editor of *Inroads,* a journal about men's soul work, and has frequently written about men's issues from a "mythopoetic" point of view.

BARTON SUTTER is a poet, fiction writer, and essayist who lives in Duluth, Minnesota. His books include *My Father's War: Stories of Midwestern Men;* a collection of essays about living in northern Minnesota, *Cold Comfort;* and three books of poems, including the most recent, *The Book of Names.*

JOHN TALLMADGE teaches literature and environmental studies in the Graduate College of Union Institute and University. He is the author of a wilderness memoir, *Meeting the Tree of Life,* and an urban sequel, *The Cincinnati Arch: Learning from Nature in the City.*

O. ALAN WELTZIEN is professor of English at the University of Montana–Western in Dillon, Montana. He has published numerous articles on western writers and is editor of *The Literary Art and Activism of Rick Bass* and coeditor of *John McPhee and the Art of Literary Nonfiction.*

TIMOTHY YOUNG, a poet and essayist who for twenty years has been associated with Minnesota Men's Conferences, lives with his wife and son in rural Wisconsin and works with juvenile offenders in the Minnesota Correctional Facility–Red Wing. He has two published chapbooks of poetry, *Men Don't Dance in America* and *Fairyland after Carnival.* His poem "The Thread of Sunlight" was included in Scribner's *The Best American Poetry of 1999.*

www.ingramcontent.com/pod-product-compliance
Lightning Source LLC
Chambersburg PA
CBHW021805220426
43662CB00006B/183